TITANIC

TITANIC

THE TRAGIC STORY OF THE
ILL-FATED OCEAN LINER

Rupert Matthews

This edition published in 2012 by Arcturus Publishing Limited
26/27 Bickels Yard, 151–153 Bermondsey Street,
London SE1 3HA

Copyright © 2012 Arcturus Publishing Limited

AD002437EN

Printed in the UK

CONTENTS

INTRODUCTION

The sinking of the RMS *Titanic* in the early hours of 15 April 1912 was the worst shipwreck the world had ever known. More than 1,500 people died when the ship went down – many from drowning but more from hypothermia on one of the coldest but most beautiful April nights anyone could remember in the North Atlantic. The sheer scale of the disaster shocked the world; the loss of life was simply horrific and utterly unprecedented.

What made the sinking seem so much worse was that it was the RMS *Titanic* that had gone down. She was the most modern, luxurious and largest passenger ship ever to put to sea. Her proud builders, Harland and Wolff, boasted that they had incorporated every known engineering feature that would make her safer. The trade press had branded her 'virtually unsinkable'. Her owners, the White Star Line, had played on her reputation for safety to run alongside the luxurious fittings of the new ship, her enormous size and her possession of the coveted status as an RMS (Royal Mail Steamer) in their sales publicity.

Yet on her maiden voyage she had gone to the bottom of the ocean with enormous loss of life.

And the lives lost were enough to make headlines by themselves. There was John Jacob Astor, one of the richest men in the world, along with fellow multi-millionaires Benjamin Guggenheim, Isidor Straus and Harry Widener. Other plutocrats, politicians and society figures were lost, each of whom would have rated a newspaper obituary in his or her own right. Also lost were the man who had designed RMS *Titanic*, Thomas Andrews, and the man who commanded her, Captain Edward Smith. In addition were the hundreds upon hundreds of more humble folk: fathers, sons, mothers, sisters and cousins. Children were left orphans, wives widowed and husbands left desolated. One entire family was wiped out by the sinking.

There was one notable and highly newsworthy

survivor in the form of Joseph Bruce Ismay, the chairman of the White Star Line which owned the *Titanic*.

Such bald statistics do not do justice to what happened that night. Unlike a modern-day train crash or aircraft disaster, the *Titanic* took hours to sink. The drama unfolded slowly and with an awful, terrifying progression. The waters crept slowly through the vast liner, pushing the huddled masses further and further towards the stern until that too slipped beneath the waves. The survivors brought with them tales of heroism and cowardice, of calmness and panic, of honour and disgrace.

The sinking of the RMS *Titanic* was a sensation around the world at the time, and has remained so ever since. There have been novels set on the doomed liner, blockbuster movies and television shows. The name is known instantly to millions and the main outline of events is familiar too.

Despite this it has proved to be remarkably difficult to pin down exactly what happened and why. Until the wreck was discovered in 1985 it had not been realized that the ship had broken in half as it sank.

In part the reasons for the confusion are natural enough. The three most senior officers on the *Titanic* were all lost, so investigators were unable to ask them questions about what had happened. Those who did survive were all too busy trying to save themselves and others to pay much attention to what strangers were doing. We know, for instance, that the *Titanic*'s band played a hymn as the ship went down, but survivors could never agree on what it was. Of particular difficulty is time. Harassed crew members and terrified passengers simply had neither the time nor the inclination to check their watches as the disaster unfolded. It is generally agreed that the ship hit the iceberg at 11.40 pm and that it finally sank at 2.20 am, but otherwise all events and incidents can be placed only approximately.

Moreover, while there was great enthusiasm to name and reward the heroes of the hour, nobody wanted to take responsibility for the disaster itself. The men who owned, operated, controlled and regulated the *Titanic* did their best to blame each other, blame the weather, blame bad luck and blame those who had died. Sorting out what had gone wrong and why was a Herculean task dogged as much by vested interests as by events.

And then there are the mysteries, the legends and the rumours. There had been a ship close to the *Titanic* when she sank, a ship that did nothing to come to the rescue. The British government put enormous effort into trying to identify this ship, but never succeeded. A great Newfoundland dog had leapt from the ship's deck and for hours had paddled beside a lifeboat. Who owned it? Had it been saved? Two toddlers unable to speak a word of English between them ended up in one lifeboat without their parents. Who were they? What should be done with them? Some lifeboats were launched when they were less than half full. For what reason had this been done?

A thousand questions surrounded the sinking of the *Titanic*. Some of them were answered, others were not. This book looks at the events of that terrible night afresh. It deals with the way the two official inquiries sought to answer the questions raised by the sinking and seeks to decide if those answers were fair and accurate in the light of new facts uncovered on the wreck and eyewitness accounts not available to the inquiries.

In the aftermath of the sinking – and ever since – there was one question above all that fascinated the world, and that more than any other needed to be answered:

Why?

CHAPTER 1

THE NORTH ATLANTIC LINERS

With hindsight it could be said that the train of events that ended with such a terrible loss of life on that April night in 1912 began as far back as October 1867, when the Royal Bank of Liverpool suddenly collapsed.

Among the many Merseyside businesses caught up in the financial turmoil that followed the failure of the bank was the White Star Line, one of Liverpool's premier shipping companies. The White Star Line had made its name running passenger and freight ships to and from Australia, but had recently begun to operate on the more prestigious and profitable New York route as well. Now, overwhelmed by debts and commitments, the directors of the White Star Line were forced to sell everything. Even that was not enough and eventually the company itself was put up for sale for the sum of £1,000. The company had nothing solid left on its books, only its name and its flag of a white star on a red, swallow-tailed pennant. As one of the best-known and most highly regarded shipping companies operating out of the Mersey River, these were worth something.

In January 1868 the name and flag were bought by an enterprising 31-year-old local man named Thomas Ismay. Young Ismay was working for the National Line and dreamed of operating his own shipping line. The purchase of the bankrupt White Star Line gave him his chance. It was a business he was born to, for he was a seaman through and through.

Ismay had been born just up the coast from Liverpool at Maryport, Cumberland, in a house overlooking the small harbour there. His father, also Thomas, was a prosperous timber merchant who owned shares in five small ships operating out of Maryport, where another member of the Ismay family ran a small shipbuilding yard. Young Thomas spent his childhood around the harbour, passing his school holidays working in his family shipyard or signed on to local ships as they plied to Ireland, Scotland and on the coastal trade. At 16 he was sent to Liverpool to be apprenticed to a shipbroker where he did well. Once the apprenticeship was over, Ismay went to sea. He served on several different ships bound for assorted ports around the Atlantic and Mediterranean. On his return to Liverpool he joined the National Line, getting his break into shipping line management through a family friend. By the time he bought the White Star Line, Ismay was a seasoned seaman and experienced shipping manager. What he lacked was money.

It was while he was relaxing one evening playing billiards that Ismay was approached by the well-known and staggeringly wealthy Liverpool merchant Gustav Schwabe. As his name might suggest, Schwabe had been born in Germany, at the great port of Hamburg, but had moved to Liverpool in his youth and by this date he had lived in the city for thirty years and had married a local girl. Schwabe was accompanied by his nephew, Gustav Wolff, who was a junior partner in the Belfast shipbuilding business Harland and Wolff. Schwabe offered to lend the White Star Line enough money to re-establish itself as a leading Liverpool shipping line and to do so on very reasonable terms. His only condition was that White Star had to agree to buy its ships exclusively from Harland and Wolff.

Understandably, Ismay did not want to lose financial control of the company he had only just bought. He therefore formed the Oceanic Steam Navigation Company which would borrow the money from Schwabe and own the ships, while the White Star Line operated the vessels. Ismay then travelled to Belfast to meet Edward Harland, the senior partner of Harland and Wolff. He found a man after his own heart.

Harland had been born in the Yorkshire port of Scarborough in 1831. Like Ismay he spent his childhood in and around ships of the coastal trade until he too was sent off to take up an apprenticeship – this time at the prestigious Stephenson engineering works at Newcastle upon Tyne. Having served out his apprenticeship learning all there was to know about steam engines, Harland moved to the Thomson shipbuilding yard in Glasgow. In 1853 he was hired to be manager of the Toward shipyard back on the Tyne, before moving to

Hickson's shipyard on Queen's Island, in the estuary of the River Lagan at Belfast.

Owner Robert Hickson was happy to leave almost every aspect of the business to Harland, who was a firm believer that the future of shipbuilding lay with iron steam ships. Harland became a notorious stickler for detail. He kept an ivory ruler in his coat pocket at all times to check every precise detail of work going on in the yard – and a piece of chalk to mark anything that did not come up to his exacting standards. He also banned smoking in the yard, considering it to be a disgusting habit and something of a danger in a workplace with so much timber lying about.

In 1858 Hickson decided to sell up and retire. He offered the entire shipbuilding yard to his manager, Harland, for £5,000. Harland did not have anything like the required sum, but one of the clerks in his office was Gustav Wolff, nephew of the wealthy Gustav Schwabe. Wolff alerted Schwabe to the business opportunity, and Schwabe contacted Harland to offer him the necessary money if he would take on Wolff as a partner. Inspired as much by Wolff's skill at designing ships as by the cash, Harland agreed. He bought the Hickson yard and so Harland and Wolff was created.

Harland's greatest contribution to shipbuilding came early in the life of the new company when he and Wolff designed the SS *Venetian* for the Bibby Line. For its day, the *Venetian* was revolutionary. The decks were of steel, not wood, and the hull was of a deep square shape in place of the V-shaped hulls of earlier vessels. The new shape effectively made the ships into immensely strong iron boxes. The V shape had been necessary in sailing ships that needed to withstand sideways pressure when the wind was on the beam, but was useless in steam ships driven from the rear by a screw propeller. The inherent strength of the box shape allowed ships to be built that were longer in proportion to their width, increasing the capacity of the holds. The increased space for cargo helped boost the profits of each voyage for the increasing numbers of lines that bought Harland and Wolff ships.

In 1862 Harland and Wolff took on a young clerk named William Pirrie. Pirrie proved to be an extremely persuasive salesman and an adept financial operator. He rose rapidly through the managerial ranks at Harland and Wolff and by 1874 he was on the board and would soon be the third partner. Increasingly the company was dominated by the three men working as a team. Harland was the inventive engineering visionary who won patents for the company, Wolff was the practical engineer who designed and built the ships, while Pirrie negotiated the orders and handled the finances.

In the 1880s, when Harland was asked how he managed his company, he replied: 'Well, Wolff designs the ships, Pirrie sells them and I smoke the firm's cigars.' He was being modest – his engineering skills were essential to the company's success.

A prosperous partnership

Over the following years, the White Star Line and Harland and Wolff saw their fortunes rise together. Both companies grew rapidly and came to dominate their respective industries. The personal affinity between Thomas Ismay and Edward Harland was cemented into a profitable business relationship. Harland and Wolff boasted to customers that they supplied all the White Star vessels, while White Star boasted to their customers that they used only Harland and Wolff ships.

At this date most merchant ships were built and operated as multi-purpose vessels. Passengers were carried on cargo ships and passenger ships carried cargo. There were two exceptions, and Harland and Wolff built both for White Star.

The speedy passenger liners built specifically for the North Atlantic run were long, slim and elegant. They had passenger accommodation divided into first class, second class and third class – White Star did not use the term 'steerage' for its cheaper cabins as Ismay felt this to be a pejorative term and valued all his customers. In June 1841, the first of these liners, SS *Columbia*, set a new record for crossing the Atlantic of 10 days and 19 hours. She thus gained for her owners, the Cunard Line, the celebrated Blue Riband, the unofficial title given to the ship that was fastest across the Atlantic. By the 1850s, Cunard was in competition with the Collins Line of New York for the Blue Riband. It was not until 1872 that White Star first gained the Blue Riband with their ship the *Adriatic*. These ships carried the prefix SS, for Steam Ship.

The second type of specialist merchant ships were the Royal Mail Steamers, which had the prefix RMS. These were ships that had won the coveted and lucrative mail contracts from the British government. These contracts were established in 1840 and linked all the major ports of the British Empire. There were many designated mail ports, of which by far the busiest was Queenstown (now Cobh) in southern Ireland, which handled most of the mail to and from Britain itself. Most of the larger British shipping lines competed for the mail contracts, including White Star and their main Atlantic rivals Cunard.

The key feature of the mail contracts as opposed to other commercial freight contracts was that there were hefty fines for being late. These varied over time, but in the 1870s when White Star entered the contest for the mail business, the amount was a guinea (one pound and one shilling) for every single minute that a mail delivery was late. While the potential profits were huge, so were the possible losses. The mail steamers did not need to be fast, though some of them were, but they did need to be reliable. They were stoutly constructed, tough ships with awesome engines and were built to cope with terrible weather.

The mail ships did not stop for anything. They ploughed on through the most violent storms without pause. They plunged their tough bows into gigantic waves that dumped hundreds of tons of water on to the decks, then lifted the stern clear of the water so that the propellers thrashed wildly in thin air before being plunged back into the waters to surge the ship forward again. The crews took to tying themselves to rails and stanchions in rough weather to save themselves being hurled into the scuppers to end up with a broken limb – or worse. Quite often mail steamers made port with smashed masts, broken rigging, shattered windows and even twisted ironworks.

Nor did the mail steamers stop for fog. Lookouts at the bows and up the mast were doubled and the officer of the watch – usually the captain in such weather – never moved from the wheel. A special fog whistle was fitted to mail steamers and in fog it was sounded for five seconds every minute to warn other vessels that a mail steamer was coming, and coming fast. Any ship hearing this distinctive warning would blow its own whistle or ring its bell to alert the mail steamer's lookouts – who were listening as much as looking. Some claimed that they smelled the air as well to pick up the distinctive odour of fishing boats or ice, both common off the east coast of Canada. As soon as a lookout called, the engines were flung into reverse and the helm put hard over to avoid whatever was looming up. Lookouts on the mail ships were specially chosen for their good eyesight, excellent hearing and quick reactions. The officers, and especially the captains, were chosen for their nerves of steel.

As well as being reliable, the mail steamers also had a second feature demanded by the government. They had to be capable of being quickly and easily converted into warships suitable to escort convoys of merchant ships. This meant that they needed to be built with gun platforms able to carry the weight of naval guns, and to have a structure strong enough to withstand the recoil of such weapons. Typically, there would be two main gun platforms, each able to carry a single gun. By the 1890s these were the cheap but reliable 4.7 in (12 cm) QF Naval Guns. One would be at the bows, the second at the stern, as these were the most stoutly constructed parts of a mail steamer. Smaller guns might be mounted elsewhere. In naval parlance these ships would become Armed Merchant Cruisers (AMCs). They were not expected to fight in battle but could prove useful in a convoy if a small enemy warship appeared.

With the need to batter their way through heavy seas, carry guns and the ever-present possibility of a collision in fog, the mail steamers were deliberately built strong. Any ship that collided with one would be guaranteed to come off worse, while piers and jetties simply crumpled. Even if a mail steamer were handled clumsily in port and bumped into a stone dock, there was rarely any real harm done. The steamer would bounce off with merely some scratched paint.

From the 1870s, White Star began to win the coveted Royal Mail contracts and so could call some of its ships RMS.

Development of the steamers

The mail steamers appealed greatly to a certain type of passenger. Businessmen and military officers often needed to arrive on time for meetings or for duty and so valued the reliability of the mail steamers. Young men liked to boast to their friends that they had ridden these Greyhounds of the Sea, as they were known; braving storm, fog and peril with dashing courage. Nobody pretended that the ships were comfortable. But they were undeniably glamorous.

Shipping lines began to realize that they could turn a healthy profit if they built mail steamers that carried passengers instead of freight. The freight holds were removed and cabins installed instead. Efforts were made to build the ships so that they were more stable in heavy seas and so more comfortable for passengers. The cabins were placed as much as possible in the centre of the ship where they would not pitch about as much, but no compromises were made regarding the tough engineering of the shipbuilding.

The culmination in this style of passenger mail steamer was widely regarded to be the White Star Line ship RMS *Oceanic*, built by Harland and Wolff. She was a ship of 17,272 gross tons, had a length of 704 ft (214 m) and a beam of 63 feet (19 m). Her engines produced 28,000 horsepower to drive her twin propellers and so speed the ship to a maximum of 19 knots, though her cruising speed was reckoned to be about 15 knots. RMS *Oceanic* could carry 410 first-class passengers, plus 300 in second class and up to 1,000 in third class. She was the first merchant ship to cost more than £1,000,000 to build and the first to be over 700 ft (213 m) long.

RMS *Oceanic* entered service in 1899 and quickly became known as the Queen of the Ocean by the men who served on her and, albeit grudgingly, by other Atlantic seamen. But even as this magnificent, tough and impressive ship took to the seas, changes were afoot that would lead to disaster.

Changes at the White Star Line

The first came in 1889 when Edward Harland retired, and he died in 1895. Then in 1892, at the age of 58, Gustav Wolff entered Parliament as the MP for Belfast East. Although he remained a partner at Harland and Wolff, he no longer spent every working day in the shipyards. At first he dropped in regularly to check on work and supervise design, but as the years passed his visits became rarer and less influential. Control of Harland and Wolff passed to William Pirrie. Pirrie was a fine businessman and a great salesman but, unlike Harland or Wolff, he was not a seaman.

The third change came in August 1899 when Thomas Ismay collapsed at his desk with chest pains. Within weeks he was dead. Control at White Star passed to his son, Joseph Bruce Ismay, who was then aged 37. Joseph Ismay had been born after his father had achieved wealth and status, so he was given the finest of upbringings. He was educated at the prestigious fee-paying school of Harrow and then set out on a tour around the world that lasted years, stopping off at universities and cultural centres approved of by his parents. After working in various shipping offices, Ismay came back to White Star in 1891 to be groomed in the arts of management that would prepare him to take over the business. Fine as his education and upbringing had been, however, it had not involved any time working at sea. Like Pirrie he was a businessman, not a seaman.

Barely was Thomas Ismay cold in his grave than Pirrie and Joseph Ismay ordered work to stop on the sister ships of RMS *Oceanic*. They had other plans for the future of the White Star Line.

Since the 1870s the passenger traffic across the Atlantic had been growing in leaps and bounds. The main expansion was in third class – or steerage as some other lines still called it. Most of this third-class traffic was one way, from east to west. Increasingly large numbers of emigrants were flooding from Europe to find a new life in North America. Most of these emigrants were from eastern or southern Europe, though some came from Scandinavia or Ireland. Apart from the Irish, few could speak English so the shipping lines that specialized in carrying them took to printing sales brochures and on-board instruction cards in a variety of languages. First- and second-class travellers, in contrast, tended to voyage back and forth across the Atlantic, as they were businessmen and their families, government officials and others travelling on business and then going back home again. And while the numbers wanting to travel in first or second class grew only steadily, the numbers clamouring for third class boomed enormously. It was into this business that Pirrie and Ismay Junior wanted to move.

They began by building what became known as the Big Four of the White Star Line: RMS *Celtic*, RMS *Cedric*, RMS *Baltic* and RMS *Adriatic*. These ships were considerably bigger than the *Olympic*. They were 730 ft (222 m) long, 75 ft (23 m) in the beam, drew 44 ft (13.4 m) of water, and weighed 24,500 gross tons. They could manage 17 knots, not as fast as the *Olympic*, but they were expected to maintain this speed throughout their voyages. Each ship could carry over 2,800 passengers, over a thousand more than the *Olympic*, most of them in third class.

Although these ships were all designated RMS and carried mail for the British government, they were not old-style mail steamers. Pirrie knew that the hefty engineering that made the older ships so strong and enabled them to carry guns was costly. There had been no wars that needed passenger steamers to convert to convoy escorts and the increasing profits from passengers more than offset the losses incurred by the occasional voyage that missed the mail deadline. The Big Four were passenger steamers pure and simple, and this was a trend that was set to continue.

Competition on the North Atlantic route

The final nail in the coffin of the old-style business built up by Thomas Ismay and Edward Harland was some time coming, but it began to be driven home in 1901. The American financier and ruthless exploiter of weakness in business competitors, J.P. Morgan, had been building up a consortium of American shipping lines that by 1900 controlled the American Line, the Red Star Line, the Atlantic Transport Line and the Leyland Line. The group was known as the International Navigation Company and kept the constituent companies operating separately to hide its true monolithic nature.

Morgan poured money into the venture with orders to the companies to embark on a price war with others operating on the North Atlantic route.

The first British company to feel the pinch was Cunard, which by 1901 was in severe financial difficulties. Morgan offered to buy Cunard, but the company's chairman Lord Inverclyde had no intention of allowing his company to fall into foreign hands. Cunard turned to the British government for help. The government agreed to give Cunard huge low-interest loans to see it through. The loans were given on condition that future Cunard ships would be designed and built with the advice of the Royal Navy. The navy foresaw that in the event of a large-scale war there would be a need for troop ships, hospital ships, ammunition ships and a host of other specialized vessels, in addition to the AMC type of converted mail steamers with gun mountings.

The first products of this partnership were the two Cunard 'superliners', the *Lusitania* and *Mauretania*. These two ships were both big and fast, as demanded by the Navy. They were 787 ft (240 m) long, 87 ft (26 m) broad and drew 34 ft (10.5 m). The revolutionary steam turbines could develop 76,000 horsepower to drive the four huge propellers, then a unique feature. They could maintain 24 knots with ease and had a top speed of 26.7 knots in short emergency bursts. The ships could each carry 552 first-class passengers, 460 second-class and 1,186 in steerage.

The events surrounding Morgan's efforts to take over Cunard had two important effects on White Star. The first was that Pirrie and Ismay concluded that they could not compete with the new Cunarders for speed. Instead they would lure passengers with comfort and luxury. It was a sensible business move. Although they would be in competition with Cunard on the lucrative North Atlantic run they would not be competing for the same passengers. Those who valued speed could book with Cunard, those who favoured comfort would book with White Star.

Morgan expands his empire

The second impact on White Star was much greater and more controversial. It was Pirrie who was the driving force. He had been both amazed and impressed by the spending power that Morgan had displayed in acquiring the American shipping lines and then launching a price war. If White Star could somehow get its hands on that vast wealth it would not only make White Star a larger shipping line, but also ensure that it would place a series of lucrative contracts with Harland and Wolff. Pirrie wanted White Star to become part of the International Navigation Company.

Ismay opposed the move, but Pirrie went behind his back to the shareholders. Not that all the shareholders were ever told the whole truth. Morgan had learned from his dealings with Lord Inverclyde at Cunard. He realized that many British people – and in particular White Star shareholders – would not want to see an important British company become foreign-owned. In 1902 Morgan changed the name of the International

Navigation Company to the International Mercantile Marine Company. Then he set up the International Navigation Company of Liverpool as a British company under British law. Pirrie was on the board of the new Liverpool company, along with a number of other British men who were ostensibly independent but were in fact paid by other companies owned by J.P. Morgan.

The Liverpool company, under Pirrie, then began buying up shares in the Oceanic Steam Navigation Company (which owned the ships operated by the White Star Line under the original deal agreed by Thomas Ismay and Gustav Schwabe back in 1868). The shares cost around £10 million to buy and the money was borrowed from a number of American companies, all owned by Morgan. The shares were then transferred to the American companies as security for the loans. To the casual observer, ownership of the shares in Oceanic Steam had merely changed from individual British shareholders to a British company. In reality, however, the company had been taken over by Morgan.

When the Liverpool company had acquired a controlling interest, Pirrie and Morgan confronted Ismay. There was nothing the unfortunate Ismay could do. If he wanted White Star to have any ships at all, he had to give in. So he did. Ismay was allowed to remain chairman of the White Star Line and Oceanic Navigation. He was even made president of the new Liverpool company as well. In reality though, he was little more than a day-to-day manager of the business. All the big, important decisions were going to be made by Morgan who acted very much on Pirrie's advice.

What Pirrie wanted was business for his shipyards, and plenty of it. Soon the orders were flowing in. Most of these new ships were unremarkable. The *Victorian*, *Armenian*, *Arabic*, *Romanic*, *Republic*, *Canopic*, *Cufic*, *Tropic* and *Gallic* were workhorse freight and passenger ships which did good service, but attracted few headlines.

Then came the SS *Laurentic*. Entering service on the Canadian run in 1909, the *Laurentic* was a ship of moderate size but revolutionary design. She was 565 ft

(172 m) in length and 67 ft (20 m) on the beam with a gross tonnage of 14,892 tons. She had first-class cabins for 230 passengers, second-class for 430 and third-class for 1,000. What made her special was her engines. The *Laurentic* had a unique mixture of traditional high-pressure triple-expansion reciprocating engines powering a pair of propellers, combined with a low-pressure turbine driving a third, central propeller. The turbine fed off the steam expelled by the reciprocating engines, providing extra power for no increase in coal consumption. She was therefore both relatively fast and cheap to run.

In 1910 the *Laurentic* made the headlines of international newspapers when Chief Inspector Drew of Scotland Yard travelled on her to overtake the slower SS *Montrose*, an older Canadian passenger liner. On board the *Montrose* was the wanted man Dr Hawley Crippen, who would later be convicted and hanged for the murder of his wife.

But the *Laurentic* and her sister ship the *Megantic*

were merely test beds for new engines and design features. What Pirrie was working towards was a vision. He wanted to cash in on the emigrant trade of third-class transatlantic journeys by building a series of huge liners that could carry vast numbers of people. The new monsters would be bigger than anything ever seen. Their great size would mean the cost of transporting each individual person would be smaller and the profits greater. Such huge ships would be possible only with massive capital investment, and only Morgan could provide that.

In 1907, Pirrie got the okay from Morgan for the massive investment that would be needed. Then he went to see Ismay in London. They met at Pirrie's London home in Belgravia. Pirrie served a fine dinner with quality wines, then as the ladies withdrew coffee and cigars came out. Pirrie broached the subject of the monster liners he was proposing. He outlined the business advantages of such big ships, their running costs and potential income. So far as we know, neither man raised nor discussed the seaworthiness of such ships. They were not seamen discussing a ship. They were businessmen discussing a business deal. Around 11 pm they reached a decision.

The *Titanic* would be built.

CHAPTER 2

LUXURY FIRST

Titanic was not to be alone. She was to be but the second of three almost identical ships: massive liners that would push the boundaries of shipbuilding to the technical limit. The first ship, *Olympic*, was laid down at Harland and Wolff on 16 December 1908; *Titanic* followed on 31 March 1909. The third ship, *Gigantic*, would have to wait until *Olympic* was completed, for there were only two slipways in the world large enough for such massive ships to be constructed and they were now both occupied.

The reason why three ships were to be built was because of the White Star's business plan for their use. Given the cruising speed envisaged for these ships – about 20 knots – the Atlantic could be crossed in good weather in about seven days. Unloading and loading the ship ready for departure would take another three days – making ten days in all. So for an individual ship it would be twenty days from leaving Southampton to being ready to leave again. With two ships on the same run, it would be ten days between sailings. With three ships on the same Southampton to New York run, the White Star Line could guarantee weekly sailings.

This would give them a big advantage over other lines when it came to wooing the paying passengers. Sailing day was set on Wednesday. Every Wednesday, winter or summer, storm or calm, one of the White Star liners was going to leave Southampton for New York. You could almost set your watch by it. Regularity, comfort and safety were what the White Star Line was promising its customers. It was a winning combination.

The sheer size of these ships is best shown by comparing them to the size of the previous record-breakers: the Big Four liners built a few years earlier. These ships, the last of which was completed in 1906, were 730 ft (222 m) long, had a beam of 75 ft (23 m), drew 44 ft (13.4 m) of water and had a tonnage of 24,500 tons. The *Titanic*, by contrast, was 882 ft (269 m) long, had a beam of 92 ft (28 m), drew 64 ft (19.5 m) and had a tonnage of 46,328 tons. She was almost twice as large as the largest of the Big Four. The bulk of that increase in size was achieved by increasing the height of the ship, though she was also longer and wider than her predecessors. There were sound business reasons for this.

The rise of Southampton

White Star intended to run the three new giants on the Southampton to New York route – the most popular and profitable route in the world. Although White Star was a Liverpool shipping line, that port had gradually become less popular for passengers. By 1905 most passengers embarking at Liverpool were from the north of England or were heading for Canada. Those from southern England or heading to the USA preferred to embark at a port in the English Channel. The most popular of these was Southampton, with its fast railway link to London.

For the shipping lines, Southampton had the advantage that a ship steaming out could call at a French port to pick up passengers from the Continent. White Star ships called at Cherbourg, which had a fast rail link to Paris. Those ships that had the coveted RMS prefix could then stop at Queenstown on the south coast of Ireland to collect the mail sacks for delivery to America. More passengers could be expected to embark at Queenstown, adding to the profits of the voyage.

The problem of this route for a ship as large as *Titanic* came at Southampton. The docks were modern, but had been constructed some twenty years earlier when ships were considerably smaller. They simply could not accommodate a ship over 800 ft (244 m) long. The three new ships meant that the Southampton Dock Company was going to have to rebuild one of their wharfs to cope, and that was expensive. Making the new ships taller rather than longer reduced the amount of expenditure required.

Almost as problematic was the entrance to and exit from Southampton. The town and its docks lay where the estuary of the River Test met that of the River Itchen to form the broader Southampton Water. This ran to the south-east to join the Solent, which then ran south-west, passing between the Isle of Wight and the mainland. Where Southampton Water met the Solent there was, and remains, an awkward corner for ships to take. Again, this put a limit on the overall length that a ship could be and still use Southampton.

With the length prescribed, the breadth was also more or less fixed. If the ships were too broad in relation to their length they would use more fuel as the engines sought to push a larger cross section through the water. If White Star was to profit from its new giant ships, the beam had to be in proportion to the length. And so it was that the *Titanic* grew upwards. It not only had more decks than did earlier liners, but it had more in proportion to its width. The *Titanic* and her sisters had eleven decks each. Any concerns that there may have been over the ships being top-heavy were met by the internal layout of the vessels.

The arrangement of these ships was complex and often confusing, but it was to have a profound effect on events that unfolded on the night the *Titanic* went down.

Among the key features of the layout was the fact that the bows and stern sections were largely the preserve of third class. This is because it is the front and rear of a ship that go up and down most as it pushes into waves, particularly the long Atlantic rollers. The central area, by contrast, rocks back and forth rather than lifting up and down. It is similarly clear that the lower decks were for second and third class while the upper decks were for first class. This was partly because the smaller cabins of second and third class necessitated more partitions and were therefore heavier and made the ship more stable by being placed lower down. Just as important was the fact that first-class passengers expected to have better access to the open decks, and these were at the top of the ship. So were the lifeboats.

The *Titanic*'s decks

At the very bottom of the ship was the Tank Top. It was here that the main heavy engineering of the ship was placed. The boilers, engines, water tanks, coal bunkers and electricity generators were all here, secured to the metal floor by massive bolts and other fittings that stopped them shifting in rough seas. The Tank Top was the tallest deck on the ship, not for the comfort of the men who toiled there but to allow space for the machinery. So huge was some of this equipment that it projected up through the decks above. The boilers and turbine engines went up through three decks, the reciprocating engines through four. These inevitably broke up the decks that they pierced, making the layout of many of the rooms there fragmented and difficult to understand.

Above the Tank Top was the Lower Orlop Deck. This was a relatively small deck which existed only in the bows of the ship and was reserved for heavy cargo.

Above the Lower Orlop was the Orlop Deck. This occupied the front quarter and rear third of the ship with the space in between taken up by boilers and engines pushing up from the Tank Top. There was no direct communication between the forward and rear Orlop. A person wanting to go from one to the other had to climb up to the Middle Deck, two floors above. The forward Orlop was occupied mostly by cargo space – the ships were expected to take on cargo as well as passengers to help boost profits. One interesting innovation on these ships – and one which White Star would promote heavily – was a cargo space especially fitted out for the transport of motor cars. There was also the capacious post room. The vast quantity of mail taken on board at Queenstown was sorted here during the voyage, ready for swift onward transit at New York, as was mail taken on at New York for offloading at Queenstown.

The rear Orlop was mostly occupied by storerooms holding food, drink and other items that would be needed on the voyage. There were large refrigerated rooms in which meat, milk and other perishable foodstuffs were stored. Two of these chilled chambers were set aside for the transport of any freight that needed to be kept chilled during the voyage.

Next up was the Lower Deck, which was more or less

on the waterline. Like the Orlop, this deck was divided between a forward Lower Deck and a rear Lower Deck with no communication between the two. The rear section was rather larger than the rear Orlop as the generators did not reach this high up.

The forward Lower Deck was largely occupied by third-class berths, together with the accommodation for the firemen, greasers and others who worked down in the engine rooms. There was also space for first- and second-class baggage that would not be needed on the voyage. Located directly above the post room was the post office. Here passengers could post letters to be delivered on arrival or lodge telegrams to be sent via the ship's radio. Next to the post office was an innovation found on no other ships: a squash court. This sport had developed during the 1850s, but had become popular in the 1890s and by the time the *Titanic* and her sister ships were being planned was the most fashionable sport among the richer bright young things on both sides of the Atlantic. The court was considered to be a major draw for richer passengers. A full-time squash professional was included in the crew to provide one-to-one coaching for those who booked lessons. A stairwell ran from the squash court up through three decks to the Saloon Deck to allow easy access from the first-class area.

The rear Lower Deck, like the rear Orlop beneath it, was mostly taken up with storerooms, some of them refrigerated. There were also a number of third-class cabins.

Above the Lower Deck was the Middle Deck, the lowest deck to run all the way through the length of the ship. Most of the forward third of this deck was occupied by third-class cabins. Behind them were various working rooms, such as the laundry. The central section of the deck was occupied by the large third-class dining saloon which ran from side to side of the ship. Behind this was accommodation for the staff who looked after third-class passengers. There were then some second-class cabins before more third-class rooms were packed into the stern area.

Next in order came the Upper Deck. This was the last deck into which the engines impinged, in the form of four fairly small blocks amidships. The bow section was occupied by third-class cabins. The central section was divided into two parts. On the port side was accommodation for crew who served first- and second-class passengers. On the starboard side were some first-class cabins that were linked by their own private staircase to the first-class areas above. To the rear of this divided section were second-class cabins and behind them more third-class rooms.

Above the Upper Deck was the Saloon Deck. The very front section was occupied by crew cabins. Behind this was a section of open deck designated the third-class promenade. It was here that third-class passengers could come up to enjoy the fresh air, play quoits or other games or lounge about on deckchairs – often then referred to as steamer chairs. Behind this came a section of first-class rooms followed by the first-class entrance hall, Grand Staircase and reception hall. The central section was occupied by the large first-class dining room, followed by the main kitchens and food preparation areas and then the second-class dining room. The rear section was occupied by more second- and third-class cabins.

The Shelter Deck above had a forecastle right in the bows that was isolated from the rest of the ship by the open air above the third-class promenade. This forecastle was filled by crew cabins, dining room and galley. Behind the open space stretched a large number of first-class cabins with, towards the rear, some smaller rooms where the personal maids and valets of first-class passengers could sleep while still being handy if they were needed. There then followed the second-class entrance, library, and a second-class promenade area. Behind these were the third-class smoking room and lounge.

The next deck up was the confusingly named Bridge Deck. The actual bridge was not here, but the deck was given this name as other steamers usually did have cabins on the same deck as the bridge and there was a certain attraction felt among first-class passengers for having a cabin close to the bridge. Nearly all of this deck was taken up by first-class cabins. Amidships were the two most luxurious suites of all. These consisted of two bedrooms, a lounge and a private promenade open to the fresh air. There was also a small adjacent cabin in which servants could sleep, as well as a first-class restaurant in which an *à la carte* meal could be ordered at any time of day – the dining room offered a set menu available only at meal times. The adjacent Café Parisien was fitted out like a French terrace café and served drinks and coffee at all hours. Behind this was another section of second-class promenade and the second-class smoking room.

Above the Bridge Deck was the Promenade Deck which, as its name suggests, was largely taken up by a wide, open promenade area that ran from one end of the ship to the other and round at front and rear. In the centre of the deck was an island superstructure in which was located a lounge, smoking room, palm court, writing room and a few cabins. This entire deck was reserved for first class.

The Promenade Deck was topped by the Boat Deck, which was mostly reserved for crew, though a short section in the centre was open to first-class promenaders. The officers' cabins were here, along with the bridge, wheelhouse and a promenade area for the use of the crew. In places the deck was blocked off for areas housing the lifeboats and the davits that were used in launching them.

Although all the decks had names, those that were used by the passengers were also given code letters. The Promenade Deck was A Deck, the Bridge Deck was B Deck and so on down to the Lower Deck which was G Deck. Contemporary accounts of the *Titanic* disaster refer to decks by both name and letter. One survivor will refer to the Saloon Deck, another to D Deck. They mean the same place.

The role of Thomas Andrews

The lead figure in the design of these three behemoths

was Thomas Andrews, the head designer at Harland and Wolff at this time. He would go on to sail on the maiden voyage of the *Titanic*. Andrews was born in 1873 into one of the more prominent Protestant families of Victorian Ireland. His father was a privy councillor and his younger brother John would become prime minister of Northern Ireland in the 1940s. Rather more relevant to Andrews' career was the fact that his mother had been born Eliza Pirrie, sister of the William Pirrie who was working at Harland and Wolff.

By 1889 Pirrie had his eye on taking over the running of Harland and Wolff as the owners, Edward Harland and Gustav Wolff, neared retirement. Pirrie was primarily a salesman, but he knew what he wanted from the company's engineering and design sections. He wanted a team that would design and build the ships that he sold to the White Star Line and to other shipping companies. He did not want men who would try to tell him the type of ships he should go and sell. As luck would have it, Pirrie's nephew, young Thomas

Andrews, was showing signs of promise as an engineer. Pirrie whisked him away from his college and enrolled him as an apprentice at Harland and Wolff.

Andrews' rise through the ranks of the company was effortless. There is no doubt that he was a talented engineer, nor that he applied himself assiduously to learning the skills of ship design and shipbuilding. He was respected by his peers and well-liked among the staff at Harland and Wolff. There was also no doubt that he owed his promotion as much to his willingness to do as his uncle told him as to his skill. In 1907, Pirrie raised Andrews to be managing director and head of the draughting department. His primary task was to be the design of the mighty trio of liners that Pirrie had sold to Morgan and to Ismay.

Andrews went to work with a will. The complex internal layout of the ships was largely his work, as were the overall dimensions. Andrews knew that the White Star Line had a reputation for luxury to uphold in its rivalry with the faster Cunard ships. Indeed, it was generally recognized that second class in a White Star ship was the equivalent of first class elsewhere. In the construction of *Olympic*, *Titanic* and *Gigantic*, Andrews set out to solidify this reputation into incontrovertible fact.

Given that the prime motivator for the construction of the ships in the first place had been the massive growth in third-class passenger numbers, it is not surprising that Andrews paid great attention to these passengers and their comfort. Unlike most other liners, third-class passengers on the three ships almost all had individual cabins holding two or four people each. Families could travel together in private, as could groups of friends who booked together. There was only one section of 'open berths', where dozens of bunks were lined up side by side. That was down in the bows of the Lower Deck and was reserved for men travelling alone. Women travelling alone were put into cabins that they shared with other single women.

Third-class accommodation was comfortable and

functional. The promenade area might have been small and squeezed in beside the forecastle with its cranes and coal-loading chutes, but at least there was one. Three hot meals were served every day in the dining room and included such hearty dishes as porridge for breakfast, bacon and potato stew for lunch and cheese with bread and pickles for supper. The provision of a lounge and smoking room was considered astonishing for third-class passengers – a type of fare-payer that other lines still dismissed as being steerage.

Second-class cabins were likewise for two or four passengers, but they had the added advantage of having built-in wardrobes and sinks with hot and cold running water. Toilets and baths were grouped together in blocks, but there were more of them than in third class so second-class passengers would rarely have had to wait to have a bath or empty the contents of the cabin's chamber pot. The public rooms of second class were carpeted – an unheard-of luxury in second class, and panelled with oak or pine.

First-class opulence

The true luxury was lavished on the first-class areas. The pampering of first-class passengers would have begun as soon as they entered the ship. While second- and third-class passengers entered into corridors to be met by stewards issuing them with instructions on how to find their rooms, the first-class passengers entered into a large entrance hall on B Deck to be met by stewards who escorted them to their rooms. These rooms were expensively panelled and were large enough to accommodate chairs and tables as well as beds. The more expensive rooms had four-poster beds and were decorated in elaborate pastiches of historic styles: Louis XIV, Jacobean, Georgian and so forth.

The public areas were fitted out to match the best that any hotel in a major city could offer. The floors were covered in the finest quality thick-piled Axminster carpets. The walls were panelled in oak or mahogany and partitions boasted fine stained-glass and filigree metalwork. Furniture was softly padded and of the best quality. Paintings by well-known artists were specially commissioned to adorn the walls. Most impressive of all was the Grand Staircase which ran up through six decks, linking all the first-class areas. It was topped by a huge glass dome that allowed daylight to flood down deep into the ship's interior.

In arranging and fitting out the passenger areas of the *Titanic* and her sister ships, Andrews and his team had been following the guidance given to them by Pirrie. They were to build a floating hotel in which every passenger was well catered for, but in which every passenger knew his or her place. The divisions between first class, second class and third class may have been complex due to the layout of the ship, but they were firm and secure. There were no doors that led directly from one section to the next. The only people intended to move between first-, second- and third-class areas were the crew, and precious few of them. Each section had its own stewards, cooks, waiters and other staff. To move from one section to the next it was necessary to go through the crew-only areas.

None of these routes was signposted, for it was thought that any member of the crew who did need to get from one part of the ship to another would know the way. On board *Titanic*, for instance, there was a staircase wide enough for several people to use at the same time that ran all the way from the Boat Deck to the Saloon Deck and another that ran from the Saloon Deck to the Orlop Deck. Both were for the use of crew only and neither was signposted as to where it went. Moreover there were additional obstructions. On older liners it was usual for the promenade decks to be open, which allowed passengers to walk about freely and, in an emergency, allowed large numbers of people to be moved from one end of the ship to the other quickly. On board *Titanic*, however, there were steel and glass screens erected at intervals to act as windbreaks to make walking the promenade more pleasant in windy weather.

In a hotel such an arrangement made sense, but on a ship it had its problems. The trend towards this sort of firm separation of the classes on board had begun some years earlier. It had come in for some criticism from sea-going officers. They pointed out that in an emergency it would be very difficult to move passengers from one part of a ship – perhaps that in which a fire was taking hold – to another and safer area. Such warnings went unheeded.

It must be admitted that during the years when liners were growing in size and their internal layouts were becoming more complex, there had been no accident in which the inability of passengers to get from one part of the ship to another quickly had been a factor. There were plenty of crew on board to show them the way if need be. In any case, all ships held boat drills during which all passengers were shepherded up to the Boat Deck, and then back down again. Thus passengers would know the way to the lifeboats, which was reckoned to be the only thing that they would really need to know.

Of course, the *Titanic* and her sister ships were a huge step up in size and complexity from the earlier ships.

Even the Cunard giants, the *Lusitania* and *Mauretania*, were 31,500 tons compared to the 46,000 tons of the *Titanic* ships. If anyone pointed this out, neither Andrews nor Pirrie took any notice.

It was not that Andrews was unaware of safety considerations when he was designing the *Olympic*, *Titanic* and *Gigantic*. He was, and took great care over them. Indeed, the measures he took impressed his contemporaries greatly.

Safety measures

The starting point for Andrews, as for any ship designer, is that they cannot stop accidents from happening but that they can design the ship so as to minimize damage if an accident does occur. Andrews will have been as aware as anyone that accidents to liners happened most often when they were in coastal waters, not on the open seas, and almost always when they were manoeuvring into or out of harbour. Those accidents came in three main forms. First the ship could run aground.

Second it might collide with something, or something might collide with it. Third the ship might catch fire. Fortunately there were established safety measures to meet all three emergencies.

To guard against running aground, all large ships were built with a double bottom. This meant that there was a second, inner watertight skin inside the bottom of the ship. If the ship ran aground and the outer hull was pierced, the inner hull would remain intact and so keep the water out. The ship would stay afloat and be able to proceed safely on its way. The damage to the outer hull could then be repaired when the ship made port. The double bottom had a secondary advantage in that it greatly strengthened the bottom of the ship. By being fixed to the steel girders to which the outer hull was also fixed, the inner hull formed a series of stiff, strong boxes.

In the *Titanic* and her sisters the double bottom was even better than that. It did not cover only the bottom of the ship but extended 7 ft (2 m) up the sides as well.

The damage caused by a collision was much the

same whether a ship itself struck something – such as a rock or another ship – or if something hit the ship. A collision typically caused a hole to be punched in the hull of the ship. Depending on where the impact occurred, this puncture might be more or less serious. A small hole above the waterline at the rear would cause little trouble. A small hole above the waterline at the bows would mean the ship would need to proceed more slowly so that waves did not break up and into the hole. Damage below the waterline was more serious. This would allow water to rush into the hull of the ship and, unless something was done, the ship would sink. In wooden ships, many of which still plied the oceans in the early 1900s, it was often possible to effect a temporary but fairly effective repair by stuffing an old sail smeared with waterproof tar into the hole – a process known as fothering. This would reduce the flow of water, allowing the pumps to discharge more water than was coming in and so allow the ship to limp to the nearest port.

Such an expedient was not possible in big liners. For a

start they did not carry sails and more importantly their hulls were much deeper in the water than any sailing ship. Even the biggest wooden ship did not draw over 20 ft (6 m), while *Titanic* had a draft of almost 35 ft (10.7 m). Fothering was effective only near the surface. Deeper down water pressure increased dramatically so either the sail would be pushed into the ship or water forced through it at a high rate.

Instead the designers of steamers had devised the concept of internal watertight compartments. These were composed of walls of iron plate that ran vertically up through the vessel, dividing the ship into a number of compartments from front to rear. The number of doors in these bulkheads was kept as low as possible, and each door had very strong hinges and locking mechanisms. If an emergency occurred, or even if a captain was unsure of his exact position and feared a collision, the doors would be shut. If the ship then hit a rock or another ship and gained a hole below the water level, the incoming water would be able to flood only the compartment where the hole occurred. Ships were designed so that no matter which compartment was pierced, the vessel would stay afloat.

Ships designed to carry passengers had long taken this concept a step further. After all, a rock or other obstruction might collide with a ship actually on the line of a watertight bulkhead, allowing water to enter the compartment on either side. Passenger ships were routinely built so that they would remain afloat with any two compartments flooded, not the one that was standard in other merchant ships.

When designing the *Titanic*, Andrews took this concept even further. He recognized that since ships were most likely to hit a rock when going forwards, the front of the ship was more likely to be holed than any other part. He therefore designed the ship so that the forward four compartments could all be pierced and the ship would stay afloat, albeit only just. Secondly he recognized that, in an emergency, the speed at which the watertight bulkhead doors could be closed would be vital. He therefore fitted each of the doors through the watertight sections with powerful electromagnets that were operated by a button on the bridge. All an officer needed to do was push the button and every door would slam shut, and stay shut as long as the ship had power.

It was an ingenious arrangement that caught the eye of the correspondent of the trade journal *The Shipbuilder and Marine Engine Builder*, who was given a tour of the ship as it was being built and who viewed the plans. After giving detailed, technical details of the watertight bulkheads and their doors the article concluded: 'The captain may, by simply moving an electric switch, instantly close the watertight doors throughout, making the vessel virtually unsinkable.'

The statement that the ships were unsinkable was never used by either Harland and Wolff or by White Star. They contented themselves with statements to the effect that every possible safety feature had been designed into the ship. Nevertheless the idea circulated that the ships were unsinkable. On the night of the tragedy this was to have a serious influence on how people behaved.

Dealing with fire was often more problematic, but by the early 20th century this was much less of a hazard than it had been. Wooden ships could burn with great ferocity and alarming speed. Steel-built liners could not burn – though their cargo and contents could. On a passenger liner it was the furnishings that were most likely to burn. To combat fire, steel liners had an ingenious system of flues. These conducted air from each part of the ship up to a pipe that emerged on to the bridge. One officer was at all times stationed to stand next to this tube. If a fire broke out, smoke would be drawn up the flues to emerge on to the bridge where it could be smelt or seen by the officer on guard. By pulling a number of levers he could isolate the air coming up the flues and so ascertain within seconds which part of the ship the smoke was coming from. A team of firefighters would then be sent down to deal with it.

In the years since this system had become standard,

only one liner had been badly damaged by fire. That was the SS *Atlantique*, a French ship. She had been steaming without passengers on board and it was suspected that the officers and crew had not been taking the same precautions as they would have done if hundreds of passengers were aboard.

The role of the lifeboats

Whether or not the *Titanic* and her sisters were 'practically unsinkable', accidents did happen, and so the ships were provided with lifeboats. Much has been made of the fact that there were not enough lifeboats on board the *Titanic* to accommodate everyone on board the ship, but in the minds of ship designers such as Andrews, this did not appear to be as foolish as it does in hindsight. The purpose of the lifeboats was not envisaged as taking off all the crew and passengers in one go and keeping them alive in mid-ocean for days on end.

We have already seen that Andrews was designing his ships to cope with the fact that any liners that did get into trouble did so in coastal waters, usually close to a major port. A liner that ran aground or was badly holed was not in imminent danger of going to the bottom. There would be an interval of hours, perhaps days, before that happened. More likely the ship would not sink at all and could be towed off for repairs. The passengers and many of the crew, meanwhile, would need to be evacuated to safety. It was assumed that other ships would soon come to the rescue. They might not want to approach too close in case they too ran aground, and so would stand off by a few hundred yards. The lifeboats would then be used to ferry the passengers and surplus crew to the rescuing ship a few at a time. If no helpful ship came by, the lifeboats could take the people ashore directly in all but the worst weather.

This, of course, did not include coping with damage that happened far from shore. A fire was the most likely such incident, and as we have seen, the *Titanic* had efficient fire protection measures. Should it prove necessary to evacuate the ship – wholly or in part

– when at sea, it was confidently expected that some other ship would come to the rescue so that lifeboats could be used to ferry the passengers and crew to safety a few at a time. After all, liners plied the busiest and most congested sea routes in the world. It was rare for them to be out of sight of another ship for very long. In any case, many ships were by this date equipped with radio. A stricken liner could summon help if no ships were within sight.

All British ships, and the White Star Line was a British company albeit one that was now American-owned, had to abide by the Board of Trade regulations. Those regulations expressed the minimum number of lifeboats in terms of cubic feet capacity of lifeboat to gross tonnage of the ship concerned. It concerned only lifeboats 'under davits' – that is, lifeboats that were attached to the launching mechanisms and could be lowered into the water at once. This might seem odd, but it did allow an inspector to quickly ascertain whether the ship met the regulations without bickering with the captain over how many people were on board or how many life rafts were elsewhere on the ship.

The problem was that the regulations had not been updated as the size of liners increased dramatically in the period from 1890 to 1910. There were no fewer than 42 categories of ship, but the largest of these was '10,000 gross tonnage and upwards'. The *Titanic* met these regulations; indeed, the number of lifeboats on board was in excess of that required. In addition, there were more boats not 'under davits' which could be used. But she was not a ship of 10,000 tons, she was one of over 46,000 tons – with a consequently much greater number of human souls on board.

In addition to the lifeboats, the Board of Trade regulations stated that passenger ships 'shall also carry approved life-belts or other similar approved articles of equal buoyancy suitable for being worn on the person, so that there may be at least one for each person on board the ship'. This the *Titanic* and her sisters had. Again, these lifebelts were designed on the assumption that another ship would soon appear on the scene. They could keep a person afloat for hours on end without the person needing to be able to swim or make any effort at all. Moreover, all lifeboats had ropes slung around their outside to which people in the water could cling. Following other wrecks, people wearing lifebelts hung on to the lifeboat ropes without difficulty for some hours before being rescued. It was accepted practice.

To summon assistance, the *Titanic* and her sisters followed standard procedure. In addition to the new-fangled radio, all merchant ships were obliged to carry the wherewithal to send out the traditional distress signals that had been used for centuries.

These were, by day, a gun fired at minute intervals, signal flags showing the letters 'NC' (standing for 'November-Charlie'), a plain square flag flying above a ball, or the continuous sounding of a whistle or foghorn. By night the traditional signals consisted of a gun fired at intervals of a minute, flames on the deck of the ship from a burning barrel of tar or similar, rockets throwing stars fired one at a time at short intervals, or the continuous sounding of a whistle or foghorn. *Titanic* carried a variation on these traditional signals in the form of Socket light signals, a type of firework manufactured by the Socket company. These consisted of a pyrotechnic charge that was launched from a mortar to a height of 800 ft (244 m) before exploding. It burst with a bang as loud as that of a gun and put forth stars similar to those of a rocket. It had the advantage that it remained viable for years, as opposed to the traditional rockets that were guaranteed for only one year. The Socket was approved by the Board of Trade, but was not in general use.

Early warning signs

It was not only in lifeboat provision that ship design had failed to keep up with the greatly increasing size of liners. Another factor was the design of the rudder. The rudder of the *Titanic* and her sisters was of a traditional design with an elegantly curved profile and the rudder post about which it rotated at the rear. This was a good design for a rudder used when the engines were creating a flow of water, whether forward or backward. The size of the rudder seems to have been calculated by Andrews and his team in relation to the length of the ship, as was then usual. However, the *Titanic* was much wider than her contemporaries. It is now recognized that the size of the rudder needs to be in proportion not to the length of the ship but to the area of water displaced by the ship at the surface. In order to be effective, the surface area of the rudder should be between 1.5 per cent and 3 per cent of the surface area of displaced water. The rudder of the *Titanic* was 1.4 per cent, putting it just outside the lower end of what today would be the recommended size range. At the time nobody thought either the shape or size of the rudder to be a design fault.

Similarly, the formation of the outer hull of ships had not changed as their size increased. There was no real reason why it should. The weight of the ship was borne by the internal girders and cross-members. The skin of the ship was there only to keep the water out. It had long been traditional to make the skin of the hull out of one-inch (2.5 cm) thick steel plate – something of a misnomer as the metal in question was more like modern iron than modern steel. The plates were fixed to each other and to the supporting internal skeleton of the ship by rivets. Again, nobody saw anything wrong in this at the time the ships were built, but as with the rudder, the increasing size of the ships was having unforeseen results.

One such result was brought into stark focus by the *Olympic* before the *Titanic* even entered service. On 14 June 1911, the huge new liner had set out on a maiden voyage from Southampton to New York that was fully booked and passed off trouble-free. Several more voyages were made without incident, until she was leaving Southampton yet again on 20 September 1911. In command was Captain Edward Smith, the most senior and most highly regarded captain working for the White Star Line. At the helm was one of the

Southampton pilots. The *Olympic* was just approaching the awkward turn into the Solent when she passed an old 8,000-ton Royal Navy cruiser, HMS *Hawke*, steaming up the Solent towards Portsmouth. What happened next surprised the commanders of both ships.

The *Hawke* was passing along the starboard side of the much larger liner when she seemed to shift sideways in the water, swinging her bows round and ramming the *Olympic* aft of the fourth funnels. The collision tore a huge hole in the side of *Olympic* and two of her interior compartments were rapidly flooded. The ship remained afloat, however, and put about to return to Southampton. There the passengers were taken off and the hull patched up so that the ship could be taken back to Harland and Wolff for proper repairs. The 'practically unsinkable' safety features had worked to perfection.

Exactly what caused the collision was more of a puzzle. The captain of HMS *Hawke* was adamant that he had not steered his ship, but that it had seemed to move of its own volition. A lengthy investigation followed.

The initial finding was that the *Olympic* had been too close to the *Hawke*, but that failed to explain how an 8,000-ton warship could apparently shift sideways in the water. The final answer came in the form of what is now termed the 'Venturi effect', named after the Italian physicist who discovered it. Put simply, when a liquid is flowing at a steady speed but then passes between two barriers, it speeds up. This in turn creates a drop in pressure.

As the two ships were passing, the tidal flow was constricted between their hulls, causing it to momentarily speed up and so cause a drop in pressure. Ordinarily such an effect would barely be noticed. But because the *Olympic* was so large it added the water it was displacing to the tidal flow, while at the same time being far too massive for the suction created to move it. The much smaller *Hawke*, however, could be affected, and was pulled violently towards the liner. The effect is now well understood and ships' captains can easily steer their vessels to compensate, but in 1911 it was all new.

There was much that was new about these vast passenger liners, and much that had to be learned. But time was running out. The *Titanic* was about to go to sea.

CHAPTER 3

THE MAIDEN VOYAGE

The *Titanic* was launched on 31 May 1911, the same day that the *Olympic* left Belfast to enter service. The double celebration attracted vast crowds to line the banks of the River Lagan and was the occasion of a formal dinner thrown by Harland and Wolff for senior figures from White Star and local dignitaries. Then the *Titanic* was moved to the nearby outfitting dock of Thomson Wharf where she would be completed.

The specification book for the *Titanic* ran to 270 pages of closely typed text, giving details of everything from the type of sinks to be fitted in cabins and the mops to be used when cleaning the decks, through to the size and shape of the masts. Some of these details were standard for all ships in the White Star Line, others were unique to the *Titanic*.

All White Star Line ships, for instance, had the same paint scheme. The hull below the waterline was to be red, the hull above the waterline to be black. The name of the ship and home port were painted in yellow. The upper works were in white with yellow detailing. The funnels were red with black tops. Every shipping line in the world had its own colour scheme, allowing seamen on other ships to recognize who owned a ship at a glance. To perform the same function at night, ships carried a variety of fireworks to let off. One line might have red rockets, another twin yellow flares. Ships passing in the night would set off these fireworks as a friendly and colourful greeting that was popular with passengers.

Stonier & Co. of Liverpool fulfilled their usual order for the crockery to be used in the dining rooms, but the *à la carte* restaurant benefitted from a special order for dinner services placed with Royal Crown Derby. Most light fittings came from N. Burt & Co. of London, but the elaborate first-class Grand Staircase had fittings supplied by Perry's of Bond Street, with the great 21-light candelabra taking pride of place.

Other more prosaic matters were organized as the ship was fitted out. Every British ship had a four-letter code that was used in official documents. Four flags, each signifying one of these letters, were hoisted by the ship as she entered and left port so that the harbour master could record the times of arrival and departure. *Titanic* was allotted the letters HVMP. The new radio system allotted a three-letter code to every official user, be it a ship, military base or private individual. *Titanic*'s radio code was to be MGY.

Throughout all this work the ship was visited at regular intervals by Francis Carruthers, the Board of Trade's ship surveyor allocated to Harland and Wolff. His job was to make sure that the ship

was constructed to the specifications already approved by the Board. He wrote dozens of reports and despite a few minor observations – all of which were rapidly dealt with by Harland and Wolff – Carruthers passed the ship as 'satisfactory' on 25 March 1912.

That meant that only the sea trials were left to complete before *Titanic* could enter service. They were scheduled for 9 am on 1 April, but poor weather delayed them until 2 April. As these were tests purely of the ship's engineering, only the engine room crew and officers were aboard in addition to those carrying out the checks. The tests were many and varied.

Staff from C.J. Smith Compasses Ltd. were on board to fine-tune the compasses. These magnetic instruments were affected by the huge amounts of metal in the ship and needed to be precisely aligned on the vessel once it was at sea and well away from other ships and dockyard cranes that could also affect them. Men from the Marconi radio company were also there to hone the radio equipment for similar reasons.

Carruthers was also on board. He was to observe the ship carrying out a number of standard manoeuvres. This included turning with the rudder put hard over while running at full speed to see how far the ship tilted. Also necessary were timed runs over set distances to test speed, raising and lowering the anchor at short notice, running with only one propeller working, turning at slow speed, going backwards, lowering and raising lifeboats, accelerating slowly and stopping with engines reversed from a speed of 18 knots. By mid-afternoon all had been finished. The ship returned to Belfast for the final paperwork to be completed. Andrews signed for Harland and Wolff to accept all the equipment, fixtures and fittings while Carruthers signed for the Board of Trade approving them all as meeting regulations. *Titanic* was ready for sea and would not need to be tested again until her annual inspection by the Board of Trade in April 1913 – an appointment she would never make.

At 8 pm *Titanic* left Belfast again, heading for Southampton where she would collect the rest of her crew, and then her passengers.

Joseph Bruce Ismay and the White Star Line had decided that their new liner would have the finest crew they could provide. They wanted everything to go smoothly on what was bound to be a newsworthy maiden voyage, and if that were not enough, Ismay himself had decided to go along for the ride. Pirrie also decided to travel on the first voyage of this mighty ship. He had missed the maiden voyage of the *Olympic* and wanted to experience the *Titanic*.

Captain Smith

The captain of the *Titanic* was Edward John Smith, known to the men who served under him as 'EJ'. He was a popular and highly regarded captain. Smith did not come from a seafaring family – his father was a pottery shop owner – but he had gone to sea in 1863 at the age of 13, as an apprentice officer with the Gibson Line of merchant ships operating out of Liverpool. His training over, Smith stayed with Gibson for some years before moving to the White Star Line in 1880. He was given the berth of fourth officer on SS *Celtic* on the Australian run, but soon transferred to the tough mail steamers on the North Atlantic. In 1887 he was given his first command, RMS *Republic*, a mail steamer of 4,000 tons.

In the years that followed, Smith gradually worked his way up the ranks of White Star captains. He commanded progressively bigger and more important ships, culminating in 1895 with his command of the *Majestic*, a 10,000-ton liner reckoned to be the most luxurious in the White Star Line. As well as serving with White Star, Smith also served with the Royal Navy Reserve (RNR). Most merchant shipping companies were happy to allow their officers leave to serve with the RNR, as they gained experience and skills that only the Navy could provide. Smith was to reach the rank of commander in the RNR, which allowed him to fly the coveted blue ensign flag on liners he commanded in place of the more usual red ensign used by merchant ships. The blue ensign was a sign of experience and skill that was recognized by passengers as well as by seamen. Some passengers would book only with a ship whose captain had the blue ensign, so such a man was a profitable asset for the company which employed him.

In 1904 Smith became commodore of the White Star Line, the most senior captain in the line. It was to him that other captains reported and he who decided which captains would command which ships. It was usual for the commodore of a line to command a new ship on its maiden voyage, and sometimes for several more voyages, before it was handed over to its regular captain.

Thus it was that Smith commanded the maiden voyages of the Big Four, as well as the *Olympic* and the *Titanic*. Smith had been on the bridge of the *Olympic* when she collided with HMS *Hawke*. He vigorously denied the allegation that his ship had steered too close to the cruiser, but later came to the view that the great engines and mighty bulk of the *Olympic* had somehow created an underwater suction.

Smith's officers

At the last minute it was decided that the chief officer of the *Olympic* should be transferred with Smith to the *Titanic* to lend his experience of the sister ship on the maiden voyage. This was Henry Wilde, one of the tallest and most muscular officers in the White Star service. Wilde was born in Liverpool in 1872 and, like Smith, went to sea as a teenager. He worked mostly for the James Chambers Line of Liverpool on merchant ships, but in 1897 he moved to the White Star Line and passenger ships. He worked his way up the ranks at a steady rate and was widely regarded as being a reliable and dependable officer who had a knack of always being in the right place at the right time when his captain needed him. The *Olympic* was his first ship as chief officer, a position he took up after spending some time ashore to deal with personal tragedy when his wife and two sons died of a sudden disease.

The arrival of Wilde on *Titanic* meant that the existing chief officer, William Murdoch, had to step down to the position of first officer. He was assured that this would be for the maiden voyage only as Wilde would then return to the *Olympic*. Murdoch's father, grandfather and three uncles were all seamen, so he had a good start when he went to sea in his turn. He gained his master's certificate at the early age of 23 and spent years on ships tramping round the world, from Shanghai to Montevideo, Auckland to New York. In 1903 he fell in love with a passenger and married her, settling in Southampton. Murdoch was noted for having a calm head in all circumstances and being able to make quick decisions.

While Murdoch was moved down to be first officer, the original first officer became second officer. This was Charles Lightoller, born in 1874, who had gone to sea at the age of 13. At the age of 20 his captain promoted him to second mate on the spot after Lightoller had managed to put out a dangerous fire in a cargo of coal. He joined White Star in 1900 with a reputation for courage and steadfast calm as well as for high-spirited practical jokes. He worked the nerve-wracking mail steamers and served under Captain Smith for a while on the RMS *Oceanic*, dubbed the Queen of the Seas. His move to *Titanic* was seen by him as a promotion in thanks for his work on other White Star ships.

The original second officer, Davey Blair, was left ashore for the maiden voyage with instructions to be on hand to rejoin *Titanic* when she returned to Southampton and Wilde left. The other officers therefore remained on board in their original positions.

The third officer was Herbert Pitman, who had been born in 1877 and went to sea at the age of 18. He had undergone special training in navigation ashore and it was largely around this tricky task that his duties on the *Titanic* revolved. Pitman was also expected to be able to relieve any of the more senior officers at any task for short periods if necessary. As third officer he was not expected to actually command the bridge for any period of time.

The position of fourth officer was taken by Joseph Boxhall. His father's family were all seamen, so at the age of 15 he too went to sea. After completing his apprenticeship, Boxhall joined the ship of which his father was captain. In 1907 he gained his extra master's certificate and, armed with this qualification, obtained a job as an officer with the White Star Line. As fourth officer, Boxhall's duties mainly involved being on the bridge to carry out whatever orders were given him by the officer of the watch. He also aided Pitman with navigation.

Harold Lowe was fifth officer. Born in 1882, Lowe had been at sea for 16 years when he joined *Titanic*.

He had spent several years on sailing ships, and had less experience on passenger steamers than the other officers. As well as being on the bridge when Boxhall was not, his duties involved liaising between the deck officers who worked the ship and the senior stewards who looked after the passengers.

The youngest of the officers was Sixth Officer James Moody, who was just 25 years old. He had been at sea since the age of 14 and was viewed as a promising young officer who was given his berth on *Titanic* to give him additional experience on passenger liners. His main duty was to stay informed about the rest of the crew, in case any men were sick or had problems that needed dealing with.

As well as the officers, there were a number of senior seamen on the *Titanic* who had specific duties. Master at Arms Joseph Bailey had the duties of a policeman, dealing with any disputes, breaking up fights and investigating any crimes that took place on board. Ship's Carpenter John Maxwell inherited his title from the days of wooden ships. He was not actually a carpenter, but an engineer responsible for monitoring the ship's fabric while at sea and effecting such minor repairs as were necessary. The two boatswains, Albert Haines and Alfred Nichols, looked after the open decks, operating the cranes, mooring ropes and winches. There were seven men with the rank of quartermaster who had duties on the bridge such as helmsman, hoisting flags, using the signal lamp and carrying messages for the officers. There were also 29 able-bodied seamen who assisted the senior seamen in their jobs. All these men had been trained in the various skills of seamanship – from tying knots to reading Morse code. They were expected to be able to turn their hand to any task involved in operating the ship.

Since the *Titanic* was carrying mail she was designated a Royal Mail Steamer. As such she had specialist lookouts, while other ships would hand the duty to whichever able-bodied seaman was to hand. There were six of these men, each chosen for their sharp eyesight and quick reactions.

Together the officers and seamen were designated as the Deck Crew.

Engineering and victualling crews

Working down below were the engineering crew who ran the engines, generators and mechanical equipment of the ship. They had various job designations such as fireman, chief fireman, electrician, trimmer, greaser and so forth. The cook and six stewards who prepared and served food to the crew were counted among the engineering crew. *Titanic* had 244 engineering crew working under Chief Engineer Joseph Bell.

Another 421 people composed the victualling crew. These included the various stewards, cooks, laundrymen and cleaners who looked after the passengers and their rooms. Of these, 21 were women tasked with looking after any single ladies travelling aboard.

In addition to the crew members employed by White Star, there were also a number of staff who worked for other companies. The *à la carte* restaurant, for instance, was run by Italian restaurateur Luigi Gatti, who owned eateries in London and on several other ships. A total of 69 staff worked in the restaurant and its kitchens.

Also on board were three British and two American post office workers tasked with sorting the mail on the voyage. The ship's band was composed of eight musicians led by Wallace Hartley, a well-known Lancashire bandmaster who worked for the musical agency C.W. & F.N. Black.

Finally there was a team of nine engineers from Harland and Wolff. It was traditional for such a group of men to go on a maiden voyage to help sort out any teething problems and to keep detailed notes on how the ship operated. The team was led by Thomas Andrews, who had designed the ship.

As *Titanic* steamed from Belfast to Southampton to prepare for her maiden voyage, Captain Smith knew that he had a problem. A coal strike had been taking place across Britain and although it was now over, there was an acute shortage of coal in Southampton. Eager to ensure that their new prestige ship would leave on time, White Star ordered other ships to offload coal to be transferred to the *Titanic*. They would have to delay their own departures until new coal stocks arrived.

This strike had two important effects on the *Titanic*. The first was that many passengers had been of the opinion that the ship would not leave on time. They had therefore cancelled their tickets and booked on a later crossing which they felt might be more reliable. The *Titanic* sailed with barely half her passenger capacity filled.

The second effect was that some of the coal loaded on to *Titanic* had not been properly damped before being poured down the huge chutes that led down into the bowels of the ship. Friction caused coal in Bunker No. 10 to catch fire. It was not a serious fire, but it was smouldering constantly. Chief Engineer Bell put his men to work to shovel out the bunker and extinguish the fire. He promised Captain Smith the fire would be out within two days. Andrews inspected the site and found the fire was not in contact with the hull, though it was heating up the plates of the watertight bulkhead between compartments five and six. Since the fire was small, would soon be out and did not affect the hull, Smith decided to put to sea on schedule.

At noon on 10 April 1912, Captain Smith ordered the Blue Peter maritime flag to be hoisted and three long blasts on the ship's whistle to sound. They were both traditional signals of imminent departure. All the passengers were on board, and so were the crew. Or at least they were supposed to be. Six firemen had gone ashore for a last drink in the Bunch of Grapes pub beside the dock entrance. At ten to twelve the men left the pub and headed towards *Titanic*. When they heard the whistle blow, two of the men ran towards the ship, but the other four did not, thinking that they still had time to get aboard. They did not get to the gangway in time and missed the ship. At the time they cursed the lost pay, but later they were to be grateful for their tardiness.

The voyage begins

The tugs took *Titanic* in hand and pulled her out of the White Star dock into Southampton Water. Once turned to face downstream, *Titanic* started her own engines and moved forward. As she passed Berth 38, *Titanic* had to negotiate a narrow passage. Tied up side by side were the liners *Oceanic* and *New York*, immobilized by lack of coal. As *Titanic* steamed by, the *New York* was caught by the same suction that had affected HMS *Hawke* as *Olympic* had gone by. The liner was yanked out of position and the cables holding her stern in place snapped. *New York* began to swing out, on course to smash into *Titanic*'s stern. But Captain Smith was not going to be caught out twice. Overruling the pilot, he put *Titanic*'s engines into full astern and brought the helm over. A collision was averted and *Titanic* went on her way.

Her first port of call was Cherbourg on the Normandy coast. Because of her size, *Titanic* could not use the docks themselves and instead anchored just off shore. Two tenders, *Traffic* and *Nomadic*, came out to meet her. They brought more passengers, luggage and freight for the giant ship to carry on her way. The sun set as *Titanic* rested at Cherbourg, while the new passengers came aboard and 24 people disembarked. At 8.10 pm, by which time it was dark, the ship's engines started up and *Titanic* again put to sea.

At 11.30 am the following morning *Titanic* dropped anchor again, this time off Queenstown. Two more tenders came out with further passengers and freight. This being Queenstown, there were also 1,385 sacks of mail destined for North America. When the tenders set off back for shore they carried with them half a dozen passengers and several dozen sacks of mail for Irish addresses. Among those leaving the ship was a stowaway. Fireman John Coffey had decided to desert and hid himself among the mail. Coffey was originally from Queenstown and had used *Titanic* as a convenient way to get home without paying a fare.

At 1.30 pm *Titanic* steamed away from Queenstown. She cruised around the southern end of Ireland then turned west, heading out into the vast Atlantic Ocean. As tradition dictated, the flag of her destination, the USA, flew from her foremast while her own blue duster – as the blue ensign is unofficially known to sailors – flew from the rear flagpole. Smith set a course that followed the standard route across the Atlantic from Britain to New York used in April. This was known as the southern route and was used by nearly all merchant ships in the winter months as it avoided the stormier weather further north and the worst of the fog and ice.

The speed Captain Smith set for his ship was about 20 knots. This was slow compared to the Cunard liners, which regularly achieved 26 knots the whole way across the Atlantic, but about average for a White Star ship. It was well known that ships' engines needed to be run in when they were new, so Smith was being prudent with his speed. Only 24 of the 29 boilers were lit.

In all, 1,343 passengers were on board *Titanic* when she set out for New York. One man who was not aboard was William Pirrie, now Lord Pirrie, the head of Harland

and Wolff. He had been taken ill and was advised not to make the transatlantic voyage.

The three passenger classes

Of the passengers on board 337 were in first class, 285 in second class and 721 in third class. They were a mixed bunch of people, but the passenger lists reveal clear and, as events would turn out, significant differences between the three classes.

The first-class passengers were almost exclusively residents of Britain, Canada or the USA. They were rich and educated people, nearly all of whom were regular passengers on liners. None had travelled on the *Titanic* before, of course, but they did know how liners operated and what to expect.

Among them were some of the richest people in the world. Colonel John Jacob Astor had an estimated fortune of about £30 million. He had been educated at Harvard and at the age of 47 was running the Astor family's vast hotel and property business, which owned over 700 different properties in New York State alone. He had fought in the Spanish-American war of 1898 and was a keen amateur engineer. With him were his new wife Madeleine, who was pregnant, and his valet Victor Robbins.

Also in first class was Benjamin Guggenheim, worth about £20 million, together with his valet and chauffeur. Isidor Straus, worth about £10 million, and his wife Ida were travelling with their manservant John Farthing. Straus was a former congressman and half-owner of Macy's department store in New York. Worth about as much as Straus was George Widener, the Philadelphia streetcar magnate.

Enjoying the comforts of the finest suite on board was Joseph Bruce Ismay, chairman of the White Star Line. Ismay had with him his valet John Fry and his secretary William Harrison.

The second-class passengers were again mostly from Britain, Canada or the USA – though there was a significantly higher proportion of British people than in first class. There were also a small number of French, South American and Spanish passengers. Again, most of them were regular travellers accustomed to liner voyages.

Third class was very different. The majority of passengers in third class were neither regular travellers, nor were they drawn from English-speaking countries. Most were emigrants moving to the USA who had never travelled by passenger ship before. A large number came from Scandinavia, where the White Star Line had a traditionally large presence. Even more came from the eastern Mediterranean and the Balkans. This was a new and lucrative market for White Star and was one of the principal reasons why the *Titanic* and her sisters had been built. A total of 113 third-class passengers were from Ireland and had boarded at Queenstown. The stewards and others who had dealings with third-class passengers encountered many language problems with their charges. But most routine issues regarding laundry, meals and directions could be sorted out by hand signals and goodwill.

On-board routine

As *Titanic* steamed to the west, those aboard settled down to shipboard routine – what there was of it. White Star prided itself on running a relaxed ship for the passengers. Food catering was quite flexible, allowing passengers a range of times to come to breakfast, lunch and supper. There was plenty of space for people to relax, smoke, write letters or play games, but nothing in the way of organized activities apart from concerts put on by the ship's band.

In contrast, the crew had a strictly regimented routine. In addition to the regular watches, the day was punctured by standard events. At 10 am each day, Captain Smith held a conference in his cabin when he received reports from the head of each section – deck crew, engineering and victualling – as well as comments any other officer wanted to make or special reports that he had asked for. This meeting usually took about half an hour, after which Smith set off on a tour of inspection. He went everywhere. Every kitchen, laundry room, boiler room, passenger lounge, bakery, open deck and saloon was visited. Only cabins did not feature on the itinerary. As was standard on White Star ships, the captain would summon the head of each department as he entered with a standard 'Anything to report?'. Hopefully the answer was 'No, sir', and Smith would move on, but members of the crew could – if they summoned up the courage – approach the captain at this point on any matter of concern. Passengers could not.

The tour of inspection ended on the bridge, where all officers were expected to be present. Smith would then discuss the inspection with his officers, confer with them over the ship's running and navigation and finally issue any orders for the coming 24 hours. It was by then usually lunchtime and Smith would go to mingle with the first-class passengers to induce goodwill and impress them with the company's solicitude for their well-being. Access to the captain was a recognized perk of paying a first-class fare. The officers used a sighting on the noonday sun to fix the ship's position. All ship's clocks were then reset to reflect that position and would remain there for the following 24 hours.

During his inspection on Saturday 13 April, Captain Smith received the welcome news that the coal fire was out, and the unwelcome news that the radio had been out of action for most of the night. There was a huge backlog of messages, so Smith ordered the radio operators, John 'Jack' Phillips and his assistant Harold Bride, to work alternate shifts to get on top of things.

There was a change to the routine on Sundays. On Sunday 14 April Captain Smith abandoned his usual tour of inspection, instead contenting himself with receiving reports in his cabin. He then led a religious service at 10.30 am in the first-class saloon for any passengers who cared to attend. The White Star prayer book contained a range of services, carefully formulated so as not to offend any Christian denomination, and consisted mostly of prayers and hymns. The service ended at 11.15 am. The passengers from second and third class who had attended were then shepherded back to their own areas of the ship by stewards.

It was usual on White Star ships to hold a boat drill immediately after religious service. On this particular Sunday, Captain Smith did not order a boat drill to take place. Nobody asked him why not. Otherwise, all was routine and normal.

At 1.42 pm, the *Titanic* received a message from another White Star Line steamer, the *Baltic*. It read 'Greek steamer *Athinai* reports passing icebergs and large quantities of field ice today in latitude 41° 51' N, longitude 49° 52' W. Wish you and *Titanic* all success. Commander.' A seaman hurried to take the message to Captain Smith and found him chatting to Ismay. Smith took the message, read it and passed it to Ismay. The ice was to the north of *Titanic*'s route, so there was nothing to worry about. Even so, Smith ordered a change of course that would take *Titanic* even further south.

It was better to be safe than sorry.

CHAPTER 4

ICE
WARNING

The ice warning that the *Titanic* received at 1.42 pm was followed at 1.45 pm by a second. This was a signal sent by the ship *Amerika* to the US Hydrographic Office in Washington DC. It read: '*Amerika* passed two large icebergs in 41° 27' N, 50° 8' W on April 14.' The message was not directed to the *Titanic* and although Bride noted it down, there is no evidence as to whether or not it was passed on to Captain Smith or to the bridge. This followed an earlier message, received at 9 am from the Cunard liner *Caronia*, which read: 'Captain *Titanic* – Westbound steamers report bergs, growlers and field ice in 42° N from 49° to 51° W. Compliments Barr' – Captain Barr being the commander of the *Caronia*.

None of these messages would have alarmed Smith or anyone else, though they would have come as something of a surprise. Ice did not usually get this far south in April. The reason was that the summer of 1911 had been exceptionally warm in the Arctic. The many glaciers that grind down to the sea on the west coast of

Greenland had shed far more icebergs than was normal. These had, as was usual and remains so today, spent the winter drifting gently around Baffin Bay before getting caught in the Labrador Current. This took them south through the Davis Strait and Labrador Sea to enter the North Atlantic and drift on south to the Grand Banks of Newfoundland.

Here the southward-flowing, icy cold Labrador Current meets the north-east-moving and much warmer Gulf Stream. The cold waters of the Labrador Current sink down beneath the warmer waters of the Gulf Stream, but the ice remains floating on the surface. It may drift about in the eddies formed by the meeting ocean streams for some while before being gently pushed off to the north-east by the North Atlantic Current. The warm water, meanwhile, is melting the ice at a considerable rate and eventually it vanishes. The upshot of this is to form a clear southern limit to the ice, a line in the ocean where there is often a higher concentration of ice than to the north. The position of this southerly limit varies from year to year, and in 1912 it was further south than normal.

Bergs, growlers and field ice

As indicated in the message from the *Caronia*, there were several different types of ice recognized by seamen at the time. The categories they devised are still in use today. The ice most often encountered in the North Atlantic is in the form of growlers, small pieces of ice rising no more than a foot or two out of the water and rarely more than a few yards across. They gained their name from the distinctive noise they made when grinding against each other. Field ice is the term used for a vast mass of floating ice composed of growlers and larger pieces of ice which might be 50 yd (46 m) or more across, but still low-lying.

Icebergs are large individual masses of ice that tower high into the air and extend deep beneath the surface. Only about 10 per cent of the iceberg can be seen above the water. In the North Atlantic, icebergs are typically steep-sided, have jagged summits and are often divided into two or more peaks. Such icebergs can become very unstable as they melt. The warmer waters of the North Atlantic Current erode the subsurface ice quicker than the cool air melts that above the surface. This can make the iceberg top-heavy, causing it to topple over suddenly and without warning.

All these forms of sea ice are composed of glacial ice which is generally clear, with a slight blue or greenish tinge. They appear to be white to the human eye because the ice surface has been pitted and scratched by waves, wind and spray. This mottled and broken surface reflects and refracts the sunlight in such a way that the ice surface appears to be white.

There was, however, a very rare type of iceberg that old-time seamen referred to as a black iceberg. This was an iceberg that had recently split, or calved, revealing a large area of ice that was completely smooth and clear. After a few days, of course, it would become pitted and so look as white as any other iceberg, but when freshly calved it showed its true nature as being clear and slightly coloured. At least it did when seen in daylight. At night, the smooth ice face appeared black and could be almost invisible.

Even so, ice was not considered to be a major hazard to shipping. No large ship had ever been known to be lost to an iceberg, though there were some suspicions. In 1901 the Alaskan coastal steamer SS *Islander* had gone down after hitting something at night – though whether she had struck an uncharted rock or an iceberg was never clear. She had limped on for some miles before sinking and the site of the collision was never established. Similarly the SS *Canadian* had sunk in 1861 off Labrador after hitting something in fog, but although there was ice in the area it was unclear if it was responsible.

Plenty of ships had been damaged by ice over the years, but these incidents had shown steam ship captains how to cope with ice rather than demonstrating its fatal dangers. Ships that hit ice tended to strike and then bounce off. If travelling at reduced speed, as in fog, the damage sustained was often nothing worse than scratched paintwork. Even if the ship was steaming at speed, the damage came in the form of a single, though at times large, puncture in the hull through which water could enter. It was, of course, to cope with this sort of damage that the internal watertight bulkheads had been developed by shipbuilders. Ships that hit ice in this way slowed down and made for the nearest port. It was a nuisance to passengers and those expecting freight – and could be an expensive business for the shipping lines that had to pay for repairs – but that was all.

Ships' captains operating in areas where ice might be expected had come to view it as just another hazard of life at sea, and not a particularly dangerous one at that. Handled properly, it was widely believed, ice was not dangerous at all. The key action to take when in areas prone to ice was to keep a sharp lookout. Seeing the ice was the most important step in avoiding damage. During the day in clear weather this was rarely a problem, but in fog or rain it could be more difficult, so extra lookouts were usually posted. Again, ice was relatively easy to spot on clear nights, as the white ice showed up well even in faint starlight. On cloudy nights there was no star or moonlight, but the waves breaking around the ice showed up clearly, even around the notorious black icebergs.

There were two ways in which steamers could cope with ice in their way. They could either reduce speed and steer a course through the floating ice or they could change course to go around it. Captain Smith had already ordered a course change that would take *Titanic* south of her planned route and well to the south of any reported ice.

That Smith's reaction was a reasonable one in the circumstances can be gauged by looking at how other captains reacted to the same ice warnings. The SS *Parisian*, an elderly liner of the Allan Line, was on a similar course to *Titanic* but some 7 hours ahead of her. She was bound from Glasgow to Boston via Halifax.

The *Parisian* had turned from a westerly course to one of west-south-west at about 2.40 pm and by 8 pm was in the longitude of the reported ice, but saw nothing as she was passing well to the south of the ice field. She was also about 15 miles (24 km) south of the route the *Titanic* would be taking. SS *Mount Temple* was a general cargo ship of the Canadian Pacific Line steaming east from New York. Captain Moore, in charge of the *Mount Temple*, was similarly steering further south than normal as a response to the ice warnings.

Rather different was the reaction of Captain Stanley Lord of the SS *Californian*. This 6,200 ton ship was a freighter and although she was fitted with cabins for passengers, there were none on board for the voyage from London to Boston. The *Californian* was on a route about 20 miles (32 km) to the north of that being followed by *Titanic* and the ship was some 4 hours ahead of the liner. Rather than turn south, Lord preferred to steam on west at his usual cruising speed of some 11 knots. At about 7 pm his lookout sighted a large iceberg and two smaller ones ahead and just to the south. The *Californian* steered around them. Lord ordered his Marconi radio operator, Cyril Evans, to send out a general message reporting the position of the icebergs.

The *Californian* takes no chances

At 10.15 pm – by which time it was dark – the lookout on the *Californian* reported seeing a faint white glow on the sea surface ahead. It was the starlight being reflected off a large amount of ice. Lord stopped his engines and swung the bows around to face north-east. In the silence that followed, Lord and his men peered west. The ice was mostly growlers and although there were some larger pieces amongst it none was large enough to qualify as an iceberg. The field of growlers was about 400 yd (366 m) from the *Californian* and stretched as far as Lord could make out to both north and south. Smaller pieces of ice bobbed about in the water around the ship. Lord concluded that the ice field was too thick to steer through in the dark and too extensive to steam around. He decided to stay where he was until dawn and then push on through the ice in daylight. He told his radio operator Evans to send out another signal giving the position and apparent size of the ice field to warn other ships.

Miles to the south-east, the *Titanic* had been making further preparations to cope with the nearby ice. At 6 pm, as night was falling, Second Officer Lightoller came on watch on the bridge. As with all steamers at night there were no lights allowed on the bridge. These would have affected the eyesight of the officers and seamen stationed there. First Officer Murdoch decided that with ice about it would be best to take extra precautions. At 7.15 pm he ordered that all lights forward of the bridge should be put out and the portholes from the crew quarters in the forecastle had their shutters secured so that no light from inside would show. At 7.30 pm the message from the *Californian* reporting the three isolated icebergs was received. The *Titanic*'s assistant radio operator, Harold Bride, took it to the bridge where it was logged. The position given was well to the north of the *Titanic*. At about the same time, Lightoller took star sightings and sent Fourth Officer Boxhall, who was assisting him on the bridge, to a lit cabin to work out the ship's position while he remained on the bridge to retain his night sight.

At 8.40 pm Lightoller sent for Ship's Carpenter Maxwell. He warned Maxwell that the sea temperature was now below the freezing point of fresh water and that precautions should be taken about the water tanks down below that were adjacent to the ship's hull. Salt water, of course, has a lower freezing point, and so the sea remained liquid. About a quarter of an hour later, Captain Smith appeared on the bridge. He had been dining with first-class passengers – Mr and Mrs Widener, the Thayer family, Mr and Mrs Carter and Major Butt. As was customary, Smith chatted to the officer of the watch, Lightoller, while his eyes accustomed themselves to the dark. Being seamen at sea they discussed the weather and how the ship was running.

The conversation then turned to the ice warnings. Lightoller remarked that it was a shame that the sea was dead calm, so that no breaking waves would show up around the base of an iceberg. Smith concurred, but pointed out the extraordinarily clear air. Both officers agreed that the lookouts would be able to spot an iceberg at some distance in the starlight. At 9.20 pm Smith headed for his cabin, located immediately behind the bridge. His last words to Lightoller were: 'If it becomes at all doubtful let me know at once. I shall be just inside.' As soon as Smith was gone, Lightoller called up to the lookouts to tell them to keep a sharp eye open for ice.

Further warnings

At 9.40 pm the SS *Mesaba* sent out an ice warning. She gave the position of the ice she had seen as 41° 25' N and 50° W. Having given her position the *Mesaba* added: 'Saw much heavy pack ice and great number large icebergs. Also field ice.' The message was received on *Titanic* by senior radio man Jack Phillips. He and Bride were still taking turn and turn about to clear the backlog of radio messages. Bride was slumped in exhausted sleep in the bed that occupied a screened-off area of the radio cabin. Phillips did not wake him to take the message to the bridge, but instead tucked it under a paperweight for later. Tragically, Phillips was not a seaman and did not understand that the position given by the *Mesaba* meant that the ice was directly in the path of the *Titanic*.

At 10 pm the watch changed. Second Officer Lightoller was replaced as officer of the watch by First Officer Murdoch. Sixth Officer Moody replaced Fourth Officer Boxhall. Up in the crow's nest, positioned part way up the forward mast, lookouts Symons and Jewell were relieved by Fred Fleet and Reginald Lee. As the new men took up their places, Lee sniffed the air and commented that ice was about. Symons agreed and passed on Lightoller's instructions. There was a short discussion about the binoculars that should have been kept in the crow's nest. These had gone missing at Southampton and had not yet been found. Chief Officer

Wilde had said that new ones would be purchased when the ship reached New York. It was usual then, as now, for lookouts to scan the horizon with their naked eyes to look for anomalies of some kind that were different from the background. Then the binoculars would be used to focus in on the object to see what it was. If a ship was spotted, the binoculars would be used to identify it, read any signals being sent, and so forth.

When the eyes of the new men had grown accustomed to the dark, the relieved men went below. As officer of the watch leaving the bridge, Lightoller set off on a tour of the ship's working areas and passenger decks to ensure that nothing obvious was amiss. Then he headed for his cabin.

At 10.30 pm the *Titanic* passed 4,000-ton SS *Rappahannock* heading in the opposite direction from Halifax to London. This small freighter with a single funnel was making about 12 knots, her usual speed. The crew of the *Rappahannock* poured up on deck to see the magnificent sight of the world's largest ship thundering through the sea at 21 knots, with light blazing from every porthole. Her captain, William Hanks, was lying ill in his cabin and it was Chief Officer Smith who had the bridge.

The *Rappahannock* had no radio, so Smith got out his Aldis lamp and signalled to the *Titanic*. Many years later, he recalled sending the message 'Have just passed several icebergs.' Seconds after Smith's signal ended, an Aldis lamp flickered from the bridge of the *Titanic*, sending the words: 'Message received. Thank you. Good night.'

A few minutes later, as Phillips continued to send the backlog of messages on to the Cape Race receiving station in Canada, Evans on the *Californian* began sending the message warning that his ship had stopped due to an impenetrable barrier of ice. From the responses to his earlier message, Evans knew that the only steamer in the immediate area with a radio set was *Titanic*, so he aimed his message at Phillips. 'Say, Old Man,' began Evans in chatty style, 'we are surrounded by ice and stopped.' But

Phillips was tired, under a heavy workload and in no mood for chat. 'Shut up, shut up,' he signalled back, 'I am working Cape Race.' For some minutes Evans listened in on the radio traffic between *Titanic* and Cape Race. After a while he realized that Phillips was in for a long session and would be unlikely to want to make contact. Evans switched off his equipment and went to bed.

Disaster is imminent

At 11.30 pm the officers and lookouts on *Titanic* noticed a very slight haze on the horizon to the west.

At 11.40 pm lookout Fred Fleet suddenly grabbed the rope linked to the alarm bell on the bridge and yanked it three times. According to the account of Quartermaster Robert Hitchens, he then shouted down the voice tube that connected to the bridge.

'Are you there?' called Fleet.

'Yes,' came the calm voice of Sixth Officer Moody. 'What do you see?'

'Iceberg. Right ahead.'

'Thank you,' replied Moody, then he turned to Murdoch, the officer of the watch.

'Iceberg right ahead, sir.'

Murdoch immediately pulled the lever ordering engineering to put the engines full astern and ordered Hitchens, who was at the wheel, to put the helm over hard to starboard. Due to the archaic helm orders that dated back to sailing days, this order meant that the bows would be turned to port. While other shipping lines had begun to adopt a new system of helm orders that more accurately reflected the reality of steering large steamers, White Star Line ships clung to the old terminology of sailing ships. These two systems of steering orders might have produced confusion among a less experienced crew, but both Murdoch and Hitchens knew their business well and it is highly unlikely that there was any error.

Murdoch's orders had been prompt and were carried out efficiently. They were exactly the orders expected of an old-style mail steamer officer on the bridge of an old-style mail steamer. They had saved dozens of ships before and the lightning reactions shown by Fleet, Moody, Murdoch and Hitchens were in the fine tradition of the mail steamers that pushed on regardless to get the mail delivered on time. But the *Titanic* was not an old-style mail steamer; she was one of Pirrie's new behemoths, and that now showed.

When the engines were put hard astern, it was only the reciprocating engines driving the outer propellers that went into reverse. The turbine driving the central propeller could not be reversed, so it was now stopped. But the rudder was directly behind the central propeller. Instead of being in the main water flow, the rudder was caught in the eddies between the flows generated by the outer propellers. It could not bite the water properly, so the ship was slower to turn than it should have been.

As the bows swung agonizingly slowly to port, Murdoch pushed the alarm button that sounded bells next to any door in a watertight bulkhead. Ten seconds later he jabbed the switch that operated the electromagnets that slammed the doors shut. The *Titanic* was now divided into a number of watertight compartments. Fourth Officer Boxhall, having heard the alarm bells, appeared on the bridge at this moment.

About 35 seconds after Fleet called the alarm, the *Titanic* hit the iceberg with her starboard bow. Other ships had hit icebergs in similar circumstances and steamed away only slightly damaged, but *Titanic* was not any other ship.

Writing afterwards, Lightoller compared what happened next to what would have happened if the ship had been one of the old mail steamers, such as the *Majestic* or the *Oceanic*. He had served on all three ships and had seen all three hit objects, so he knew what he was talking about. He believed that if the *Majestic* and perhaps the *Oceanic* had hit the iceberg a glancing blow, as the *Titanic* did, they would have been thrown sideways by the impact and escaped with a single hole in one watertight compartment.

The *Titanic*, however, behaved very differently.

Whereas the *Majestic* had been a ship of 10,000 tons that cruised at some 17 knots and the *Oceanic* one of 17,000 tons cruising at 18 knots, the *Titanic* was a massive 46,000 tons doing 22.5 knots when she struck. Momentum is a function of mass and speed, so *Titanic* had a momentum six times that of the *Majestic* and 3.4 times that of the *Oceanic*. Yet her hull plates were of identical strength and thickness.

The blow struck on the starboard bow of *Titanic* was simply not enough to push her sideways because she had too much momentum. Instead the inherent give or flexibility in the ship's hull allowed the bows alone to bend slightly sideways away from the iceberg. Then the bows sprang back again to strike the iceberg again and inflict a second hole. Again the bows were pushed sideways and again came back to hit the ice and sustain another wound. And again. And again. And again. While the *Titanic* ploughed straight on.

In all, the first five watertight compartments had been breached. None of the holes was particularly large, as was usual with iceberg strikes, but they were all letting in water. The ship could have continued on her voyage with two watertight compartments flooded, perhaps with three, and would have stayed afloat to be towed to harbour with four breached. But with five flooding it was inevitable that she would go to the bottom.

Although nobody on board yet knew it, the *Titanic* was doomed.

CHAPTER 5

WOMEN AND CHILDREN FIRST

The impact as the *Titanic* bounced along the side of the iceberg, tearing open her watertight compartments, was slight. Most passengers did not even notice it as they were by then asleep in their cabins, as were many of the crew.

Even those who knew what had happened were not too concerned. Lookout Frederick Fleet watched the iceberg drift silently past him on the starboard side. 'There was just a slight grinding noise,' he reported later. 'I thought it was a narrow shave. Some ice fell on the forecastle and on the weatherdeck [the third-class promenade], but not much only where she rubbed against it.' He and lookout Reginald Lee went back to their job of watching ahead and around the ship.

One passenger who was awake was French cotton trader Alfred Omont. He was sitting up in the Café Parisien playing cards with two other Frenchmen and an American they had met on board. Omont recalled: 'We played on and then there was a shock. I have crossed the Atlantic thirteen times, and the shock was not a great one, and I thought it was caused by a wave. After about a few minutes I asked the waiter to put down the porthole and he did so. And we saw nothing. When the shock had happened, we saw something white through the porthole and we saw water on the ports. When the waiter opened the porthole we saw nothing but a clear night.'

Also awake was Second Officer Charles Lightoller, who had returned to his cabin after doing the rounds. He later said,

I had just switched the light out. I was going to sleep. I had switched the light out and turned over to go to sleep, but I was awake. There was a slight jar, followed by this grinding sound for a couple of seconds. It struck me we had struck something and then thinking it over it was a feeling as if she may have hit some obstruction with her propeller and stripped the blades off. There was a slight bumping. I lay there for a few minutes, then feeling the engines had stopped I got up. I went on to the deck on the port side, my cabin is on the port side. Everything was normal. I crossed over to the starboard side and could see the commander [Captain Smith] standing on the bridge.

Lightoller then returned to his cabin, reasoning that if he were needed he ought to be where people could find him easily.

Down on C Deck in the first-class cabin C51, retired US army officer and historian Colonel Archibald Gracie was fast asleep. Gracie recalled later that:

I was aroused by a sudden shock and noise forward on the starboard side, which I at once concluded was caused by a collision, with some other ship perhaps. I jumped from my bed, turned on the electric light, glanced at my watch nearby on the dresser, which I had changed to agree with ship's time. It was 11.45. I opened the door of my cabin, looked out into the corridor, but could not see or hear anyone – there was no commotion whatever.

The few other passengers who bothered to look found much the same thing. There was the usual quiet stillness of a liner at night.

There was plenty of commotion going on down in Section 6 of engineering, near the bows on the Tank Top Deck. Fireman Fred Barrett was talking to Second Engineer James Hesketh when, according to Barrett,

The bell rang, the red light showed. We sang out that the doors were shutting and there was a crash just as we sung out. The water came through the ship's side. It was a large volume of water about 2 feet [60 cm] above the floor plates. The engineer and I jumped to the next section. The next section to the forward section is No. 5. I went back to No. 6 fireroom and there was 8 feet [2.4 m] of water in there. I went to No. 5 fireroom when the lights went out. I was sent to find lamps as the lights were out, and when we got the lamps we looked at the boilers and there was

no water in them. I ran to the engineer and he told
me to get some firemen down to draw the fire. I got
15 men down.

There was good reason for the rush of men down to the engine rooms. The boilers were huge constructions filled with steam. If cold seawater struck them the steam inside would condense back to water, creating a huge and powerful vacuum that would cause the boilers to implode violently. Not only would this destroy the engines, it could cause other serious damage to the ship.

Assessing the damage

Things were rather calmer, but no less dramatic, up on the bridge. Barely had the grinding sound ended than Captain Smith was there. Hitchens later recalled the conversation:

'Mr Murdoch,' Smith called to the first officer who was on watch, 'What was that?'

'An iceberg, sir,' replied Murdoch. 'I hard a starboarded

and reversed the engines, and I was going to hard a port around it, but she was too close. I couldn't do any more.'

Smith then ordered Murdoch to close the watertight doors, which was already done, and strode out on to the open-air section of the bridge, presumably to try to see the iceberg. This was where Lightoller saw him.

Although Fourth Officer Boxhall was off duty, he had not gone to bed and had raced up to the bridge when he realized something was wrong. He came out on to the open section with the captain. Smith sent him down below to look for any damage to the ship's hull. Boxhall recalled:

I went right down below in the lowest steerage in the forward part of the ship as far as I could possibly get without going into the cargo portion of the ship [this would have been Lower or G Deck] and inspected all the decks as I came up in the vicinity of where I thought she had struck. I found no damage, I found no indication that the ship had damaged herself.

Then I went on to the bridge and reported that I could find no damage.

Captain Smith was not deceived. He had been at sea too long to believe that *Titanic* had escaped without any damage at all. He sent for John Hutchinson, the ship's carpenter, with orders to sound the ship; that is to check if any water was in the bottom of the ship or in the false bottom below the Tank Top Deck. But Hutchinson knew his job well; he had already done this and was hurrying up to the bridge with the bad news. He reported that water was entering the forward part of the ship rapidly. Smith sent him off again to carry out a more detailed survey.

Chief Officer Wilde also knew his job. As soon as he heard the collision he hurried down to the crew quarters at the very front of the Lower Deck. There he found Boatswain Albert Haines, Lamp Trimmer Samuel Hemming and others standing by the iron wall that separated the crew quarters from the chain locker, a large chamber in which the iron anchor chain was stowed when the anchor was lifted. He asked them what was going on. Haines told him that air was hissing through an overflow pipe from the chain locker.

'I said the forepeak tank was filling,' Haines later recalled. 'The air was coming out and the water was coming in. He [Wilde] asked if there was any water in the forepeak [a lockable storeroom for the use of the crew], and the storekeeper went into the forepeak and there was no water there, but the forepeak tank was filling. The chief officer then went on to the bridge to report.' Haines was not sure, but he thought that this happened at about 11.45 pm.

Meanwhile, Smith sent another seaman to rouse Thomas Andrews, the Harland and Wolff engineer who had designed *Titanic*. He and a team of eight other Harland and Wolff men were on board to monitor how the ship handled, deal with any teething problems and generally assist the crew with the new ship. They would be desperately needed in the hours that followed.

Andrews was found awake and fully dressed in his cabin, catching up on paperwork. He had not noticed the impact at all and was surprised to get a message summoning him to the bridge. Nevertheless he put down his pen, slipped on a jacket and set off.

Another man had been sent to get Joseph Bruce Ismay, head of the White Star Line. He was in bed, but was quickly roused. He hastily pulled trousers and a coat over his pyjamas and headed for the bridge. The American historian Colonel Gracie had by this time wandered up to A Deck to find out what was going on. He recalled that:

> [Upon] entering the companionway, I passed Mr Ismay with a member of the crew hurrying up the stairway. He wore a day suit, and, as usual, was hatless. He seemed too much preoccupied to notice anyone. Therefore I did not speak to him, but regarded his face very closely perchance to learn from his manner how serious the accident might be. It

occurred to me then that he was putting on as brave a face as possible.

In fact at this point Ismay knew little more than Gracie.

While Smith's messengers were rousing Ismay and Andrews, a Cornish mail clerk named Jago Smith had appeared on the bridge. He had been in the mailroom and came to report that water was coming in among the sacks of mail and was already 2 ft (60 cm) deep. This was bad news indeed. The mailroom was on the Orlop Deck and some distance back from the bows. If that was flooding, the damage must be fairly serious.

Captain Smith left the bridge and walked back along the Boat Deck to the radio room. The two Marconi radio operators were still hard at work trying to clear the backlog of personal messages that had built up while the radio was out of action, and keeping on top of the routine ship and navigation messages. Jack Phillips, the senior man, was due to be on duty from 10 pm to 2 am, and his assistant Harold Bride had been sleeping in a bunk in a screened-off section of the radio room. Bride had been woken up by the bustle of men racing to and from the bridge and had got dressed. He felt refreshed and told Phillips to take a break. Phillips was taking his boots off ready to sleep fully dressed when Smith entered.

'We have struck an iceberg,' announced the captain. 'I am having an inspection made to see what has been done to us. You'd better get ready to send out a call for assistance, but don't send it until I tell you.' He then headed back to the bridge.

Ismay arrived on the bridge next and asked Smith: 'Do you think the ship is seriously damaged?'

Smith paused for a moment, and then replied: 'I am afraid that she is.'

'How long has she got?'

Andrews arrived a few seconds later. He and Smith left the bridge to go below and make their own inspection of the rising water. They found the mailroom now under 3 ft (90 cm) of water, before moving on to the engineering section. Fourth Officer Boxhall had decided to go back below and arrived in the mailroom just after Smith and Andrews left. 'I looked through the open door and saw mail clerks working at the racks, taking letters out of the racks, and sacks of mail floating about.'

He then turned to go back up to the bridge, reasoning that he would be needed there. As he clambered up through the decks he passed increasing numbers of passengers and crew in the corridors. Third-class passengers from the lowest cabins in the bows seem to have been the first to move – more of them had been woken by the impact as they were closest to it. And the crash had been followed by the noise of engineers and other crew members clattering about. Up in first- and second-class areas both men and women were in the corridors wondering what was happening and hearing rumours of a brush with an iceberg.

At about midnight, Smith and Andrews returned to

the bridge. Andrews grabbed some paper and a pencil and did some rapid calculations.

'How long has she got?' asked Captain Smith.

Andrews looked up from his paper with hollow eyes. 'An hour and a half, or two hours. Maybe a little more, but not much.' It was the first indication anyone had that the ship was sinking. After all, the *Titanic* had been built with every conceivable safety feature. The trade press had described her as 'virtually unsinkable', and the popular press had repeated the claim time and again. Even experienced seagoing officers thought that it would take a lot to sink the *Titanic*, and ice was not considered to be a serious danger. Now they were learning differently.

Suddenly, Captain Smith was all action. He sent Chief Officer Wilde, who by now was on the bridge, to get the covers off the lifeboats. All the lifeboats were kept securely covered by heavy tarpaulins to protect them against the North Atlantic weather. First Officer Murdoch was sent to rouse the stewards and tell them to wake up the passengers and to get them all dressed.

Fourth Officer Boxhall was sent to find Second Officer Lightoller and Third Officer Pitman, while Sixth Officer Moody was packed off to find the written plans for evacuating the ship and summoning assistance.

The evacuation plan was to prove a problem throughout the sinking of the ship. In theory every ship had an evacuation plan, which specified what each individual member of the crew was supposed to do in the event of an emergency, or where he was to report for duty if for some reason he could not perform his allotted task. The entire crew was supposed to be conversant with their particular roles in the evacuation plan. Although each crew member had been given these instructions, the *Titanic* was a new ship with a new crew. Many of the stewards were still trying to find the best route from the rooms they cared for to the kitchens, to the toilets, to the baths, to their own sleeping quarters and back again. None of the engineering crew had found the time to go up to the Boat Deck, and few on board knew which lifeboat was which.

The unprecedented size of the *Titanic* and the complex internal layout of the ship would also cause problems. As events unfolded this would prove to be particularly difficult for those in third class. They were deepest in the bowels of the ship and furthest away from the lifeboats. Some of them were faced with a walk of over a quarter of a mile (400 m), and a climb of seven storeys, to get to the Boat Deck – and that was if they knew the way. Of course, they did not know the way. The rigid delineations between the accommodation areas for the three classes of passenger meant that third-class passengers simply did not go to the Boat Deck, so in an emergency they did not know where it was. Neither did second-class passengers for that matter, but they were generally closer to the boats and could work out how to get there more easily.

As with so much about the *Titanic*, the evacuation plan was a traditional one that failed in the novel environment of the huge liner. The plan was based on those that had worked well enough in liners of 15,000 tons, but in a liner of 46,000 tons it failed. The ship was simply too big and too complex for the old-style evacuation systems to function properly.

It was a recipe for confusion, but fortunately *Titanic*'s crew was composed of highly experienced men and women. If they did not know how to reach their precise emergency position, they did at least know what sorts of things they were supposed to do. By and large, they all found something useful to do. Discipline also proved to be strong among the crew and they responded well to instructions. Officers, senior stewards, senior engineers and others responsible for particular tasks all found themselves rounding up crewmen and issuing orders. Without exception, those orders were obeyed.

The distress signals begin

Having heard the news that the *Titanic* was sinking, Ismay wandered off the bridge, apparently in shock. Nobody took much notice of him as they were all too busy attending to their tasks. It was, however, the first of some odd things that Ismay did that night.

Meanwhile, Smith strode briskly to the radio room. He put his head around the door.

'Send the call for assistance,' he said curtly and began to leave.

'Now?' queried Phillips.

'Yes. At once,' snapped Smith, and returned to the bridge. Noticeably, Smith had not told Phillips and Bride that the ship was sinking, only that they needed assistance.

It was 12.15 am when this first distress signal went out. Conforming to the standard procedures of the day, Phillips sent the letters CQD, the three-letter code that indicated a call for immediate assistance. This was followed by the code for the *Titanic*, MGY. A few minutes later, Fourth Officer Boxhall came in with a piece of paper on which was written the position that he had worked out for the *Titanic*. Phillips added that to the signal and continued sending. Each time he completed the call for help, he would pause to wait for an answer. The minutes ticked by. No answer came. Nobody, it seemed, had heard.

Meanwhile, Second Officer Lightoller had been roused by Fourth Officer Boxhall. He was heading for the bridge when he met Chief Officer Wilde and First Officer Murdoch. Wilde told him to start clearing the lifeboats on the port side. Wilde himself went off with Murdoch to clear the lifeboats on the starboard side. Very soon after that the deckhands appeared with Third Officer Pitman. Lightoller divided them into two groups, keeping Pitman and half the men for himself and sending the rest over to the starboard side.

At this point in time, neither Lightoller nor Boxhall knew that the ship was sinking. That knowledge seems to have been kept to Captain Smith, Chief Officer Wilde, Andrews and Ismay. Quite why this was so has been a matter of dispute. It is most likely that there were two reasons in play at this point.

The first was that Smith was very busy issuing orders, checking progress and monitoring what was going on. He was probably so busy that it may not have occurred to him to pass on extraneous information. The lifeboats had to be readied, the passengers had to be prepared – so he may have thought it was enough that the crew were given the right orders.

Second, Smith probably shared the very real fears that Lightoller, Boxhall and others say that they felt later on, as the situation became clearer to them. That was the fear of panic. Evacuating a ship at sea is a complex and delicate task. Even if another ship is standing by to take off the passengers and crew, there can be problems. The passengers need to stand by while lifeboats are prepared, then wait their turn to get a place. Once in the lifeboat, they have to sit quietly and still while the crew lower the boat and row it away. It does not take much to overturn a lifeboat. The passengers must then disembark from the lifeboat on to the rescue ship as quickly as possible so that the boat can go back for more people.

A panicking mob of hundreds of passengers would be a real problem. People would be trampled underfoot as the crowd surged up the stairs to the Boat Deck. Others might be shoved overboard into the water. Lifeboats could be overturned and lost. Even if nobody was killed or badly injured, panic would cause delays and slow down the evacuation. It was to be avoided at all costs. This much was true of any ship at sea that got into difficulties.

But the *Titanic* had some very special problems of its own on that freezing cold, beautifully starlit night. For a start there was no rescue ship. The liner was in the middle of the main North Atlantic sea lane. There should have been half a dozen ships within sight of her, but because of the coal strike those ships were tied up in port. There were very few boats about in the North Atlantic. Moreover there was, as yet, no response to the radio distress signal.

That meant that there were only the lifeboats on board *Titanic* to help her stricken passengers. Given the bitterly cold temperature of the water it was unlikely that anyone who fell into it would survive for very long. In warmer seas people had bobbed about for hours without any ill effects, but that would not happen this night. Only those in lifeboats would stand any chance of survival. And there were simply not enough lifeboats. The boats could between them hold 1,178 people, but there were 2,223 souls on board. Once this became generally appreciated, some form of panic was almost inevitable.

Lightoller, Boxhall and others would later have another worry to contend with. If the problem of getting people into the lifeboats without panic was not difficult enough already, there was an added complication when it came to the third-class passengers. Most of them could not speak English. They would be quite unable to understand the instructions and orders given them by the crew. That alone would cause delays and confusion, but even worse was the firm belief among the British crew that people from the Mediterranean – particularly Italians and Greeks – were much more excitable, nervous and prone to panic than were British, American or German people. The junior officers, and perhaps Smith also, believed that if those travelling in third class knew the true situation there would be a mass panic. Lifeboats would be overturned and the death toll even higher than it would otherwise be. Whether the Italians and Greeks actually were more excitable than the Britons is beside the point. British sea officers believed it to be the case, and so they acted on that assumption.

The Birkenhead Drill

All seamen understood the problems that panic could cause. They knew also that in such circumstances it was the weakest who suffered the most, and on passenger liners that meant women and children. It was for this reason that the custom had grown up at sea that in any hazardous situation it was a case of 'women and children first'. Always it was women and children who got lifebelts first, women and children who were evacuated first, and women and children who got into lifeboats first. The men, it was expected, could take their chances at swimming. After all, men were physically stronger and stood more chance in the sea than would a woman, still more so than a child.

This custom was often termed the Birkenhead Drill, after the wreck of HMS *Birkenhead* in 1845. That ship had been transporting 550 men of the 73rd Regiment to Algoa Bay in South Africa, together with seven women and 13 children. The crew totalled 80 men. At 2 am on 24 February the ship hit an uncharted rock near Cape Town. Within minutes it was clear she was doomed. The captain, Robert Salmond, ordered the ship's boats to ferry the passengers and crew to the coast that lay only 2 miles (3 km) away. The army officer in command of the regiment, Colonel Seton, ordered his men to draw up on deck as if on parade. The serried ranks of soldiers stood stiffly to attention while the women and children were loaded on to the boats, then the men were marched off a few at a time until the boats were full. The boats, rowed by sailors, began pulling to the shore.

Suddenly the *Birkenhead* broke in two and sank, pitching all those left on board into the waters. Some clung to the wreckage, others began swimming to shore. Sharks closed in and began to feed off the living and the dead. By the time a passing schooner arrived to effect a rescue 450 had died – but all the women and children had been saved.

Like all seamen, Captain Smith would have had the example of the *Birkenhead* in his mind as he received the news that his ship was doomed. His primary duty was to get the women and children off in lifeboats first, and then to allow such men as there was room for to follow. To achieve that, panic must be avoided at all costs. If that meant not telling even his own officers that the *Titanic* was sinking, then so be it. They would realize all in good time.

Furthermore, Smith knew that it was his personal duty to make sure that as many passengers as possible got off safely. He himself would be the last to leave. Given the coldness of the waters and the lack of lifeboats, that meant certain death unless a rescue ship appeared quickly. Smith certainly had a lot to think about as the *Titanic* settled beneath his feet. It was a little past midnight when he realized the ship was lost and about a quarter past midnight when the lifeboats began to be uncovered. *Titanic* had barely two hours left to live.

In the meantime, other senior crew had been summoned to the bridge to be given their instructions. Among these were Chief Steward Andrew Latimer and Assistant Purser Reginald Barker – for some reason nobody could find the Chief Purser Hugh McElroy. They took away the orders for the victualling crew. Chief Engineer Bell had not come to the bridge as he was busy in the boiler rooms trying to assess the damage. A junior engineer had come in his place and was sent racing back down with Smith's orders to Bell. These were to draw the fires; that is to put out the fires under the boilers and to cool the boilers down. However, sufficient energy was to be kept going to power the electricity generators. It was essential that the lights remained on throughout the ship and that the radio room had enough power to send out distress signals.

Bell sent his assistant engineers to carry the messages through the engine rooms. Nobody in the engine rooms can have been in much doubt as to what the orders meant. If the fires were drawn, it meant the engines were not going to be needed again. It began to dawn on the engineering crew that the ship might be sinking.

Not only were the fires drawn, but the high pressure steam had to be released – and quickly. All the outlets were opened at once, allowing the steam to pour out of the engines. This produced a terrific noise that lasted for over a quarter of an hour. Up on the open decks it was only possible for men to make themselves understood by shouting into each other's ears. Down below the noise was not quite so bad, but everyone heard it and wondered what it meant. It was about 12.20 am.

The senior stewards had already been summoned to a meeting with Chief Steward Latimer and Assistant Purser Barker to be given their instructions. Latimer himself would look after the first-class areas. Chief Second Class Steward John Hardy was to be in personal charge of the second-class passengers and Chief Third Class Steward James Kieran was to look after passengers in third class.

Each of the cabin stewards was already responsible for a particular set of rooms during the voyage. He was to answer any queries from passengers in those rooms, call them for meals, be on hand to run errands and so forth. In first class each cabin steward looked after just eight cabins. He was expected to know the names of all the passengers he cared for and to be familiar with their individual foibles and desires. Second-class cabin stewards looked after about twice as many passengers, while those in third class catered to twice as many again and were not required to know passengers by name or to treat them as individuals. There were many other stewards, classed as deck stewards or saloon stewards, working as waiters in the dining rooms, serving drinks in the bars, organizing the library, operating the lifts or performing any one of a dozen other jobs.

In this emergency the first and most important task was for the stewardesses to locate the single women travelling alone for whom they were responsible. They were to get these women dressed, make them don their lifebelts and take them up to the Boat Deck at once. It would seem that some of the deck stewards were sent off with the stewardesses to help with this task.

'You do not have much time'

The first-class stewardesses, Mrs Katherine Gold and Mrs Annie Martin, experienced some difficulty in persuading the ladies travelling alone to get dressed, put on their lifebelts and get up on deck. They managed to get their charges as far as the Grand Staircase, but that was as far as they got. Some of the ladies protested that going out into the freezing night air would be bad for their health, or that of their children, while younger women were larking about and refusing to put on their lifebelts. The knot of women was suddenly stilled by the booming voice of Thomas Andrews, designer of the *Titanic*.

'You must hurry up on deck at once,' he said. 'You do not have much time.' There was something about the way he spoke that ended all the jokes and capers and silenced those objecting to the cold. The group of women trooped up to the Boat Deck.

Next, each of the cabin stewards was given the task of waking up the passengers in the cabins that they cared for. The passengers were to be told to dress, get into lifebelts and then wait with their steward in the public areas for further orders. As each cabin was emptied, the steward locked it behind its occupant. It was not unknown for less honest passengers to take advantage of boat drills or of emergencies to steal jewellery or money from cabins. The White Star Line had a firm policy of locking all cabins during drills or emergencies once the occupants had been evacuated. This policy was not always rigorously enforced in third class – after all, there was less to steal from those cabins.

The lifebelts were stored in the cabins themselves, one for each passenger. They were shaped rather like a waistcoat or vest and were thickly padded with cork. They were capable of keeping an adult afloat even if the person in question was unable to swim or was unconscious. There was more cork in the front of the lifebelt than the rear to ensure that the face was kept out of the water. However they were rather clumsy garments and there were to be many delays as passengers tried them on first under overcoats, then on top of them. They also made each passenger rather fatter than normal, which caused problems squeezing through doorways and up ladders.

According to Chief Second Class Steward Hardy:

Immediately after the collision I sent for all stewards. Purser Barker gave the order to put on lifebelts. The stewards were interested in their own cabins. They all came along and I went among the people and told those people to go on deck with their lifebelts on and we assisted the ladies with the belts and we assisted in getting the children out of bed. The whole of the men stewards came and they assisted me in going around calling the different passengers. I got them all up on the outer decks and they were grouped about the ship

in different parts. And I went up to my station on
the Boat Deck, Boat 1 on the starboard side. It was
about 12.30. I spoke to [First Officer] Mr Murdoch.
He said, 'I believe she is gone.'

Once the second-class ladies travelling alone were up
on deck, the matron who was in charge of children in
second class, Mrs Catherine Wallis, did a quick tour
round her charges to make sure they had warm clothing
on. Then she touched a steward on the arm. 'I am going
back to my cabin where I am safe,' she said. She was
never seen again.

Down in third class, Steward John Hart was roused
from his bed and told to hurry up and get dressed. He
recalled later:

It was the chief third-class steward, Mr Kieran. He
passed several orders. To me he said 'Go along to
your rooms and get your people about. Get lifebelts
placed on them, see that they have lifebelts.' That was

Section K on E Deck [towards the stern], I had 58 passengers, no single men and nine married couples with children. I went to each third-class room and roused them. The majority were up. I saw the lifebelts placed on them [sic] that were willing to have them put on. Some refused to put them on. They said they saw no occasion for putting them on. They did not believe the ship was hurt in any way. Mr Kieran called 'Have you placed lifebelts on those willing to have them?' I said 'Yes'. He said, 'There will be further instructions. Stand by your own people.'

Hart did his best to keep the third-class passengers from his cabins standing quietly and all together, ready for the next orders. This was not proving easy. Increasing numbers of other third-class passengers from lower decks and from near the bows were coming up to E Deck. Several of them had come without lifebelts and, seeing those from E Deck wearing theirs, wanted lifebelts as well. Added to the growing numbers and demands for lifebelts was the confusion caused by the language problem. *Titanic* had both an Italian and a Hungarian speaker among the third-class crew to act as translators, but Hart could see neither of them.

Roused along with the stewards, but not given any definite task to perform, was the ship's band. Led by bandmaster Wallace Hartley, the musicians gathered their instruments and climbed up to the large first-class lounge on A Deck, immediately beneath the Boat Deck. They struck up some lively dance tunes to keep the passengers entertained and raise spirits.

Also woken, but with no specific role in the emergency drill, was the ship's chief baker, Charles Joughin. He was, however, an old hand on liners and had been cooking at sea for over ten years. He knew that if passengers were put off in lifeboats they could well be bobbing about for some hours before the ship was deemed safe enough for them to return. They might get hungry. Joughin rounded up his thirteen bakers – a baker's dozen of men – and told them to go through the kitchens to grab as much food as they could. This was then loaded into wooden boxes and barrels to protect it against splashing seawater. Joughin then led his team up to the Boat Deck and supervised the loading of provisions into the boats.

Task done, Joughin reasoned he would not be wanted again for some hours so he slipped back to his cabin and poured himself a stiff whisky. Joughin was a notoriously heavy drinker – though he never allowed himself to be drunk on duty – so he kept his job when other drinkers lost theirs. On one ship he was found to have a fully working still in his cabin, for which he received the only reprimand in his long career at sea.

Up in first class, Mrs Edith Graham, her teenage daughter Margaret, and the girl's governess Elizabeth Shutes, were still fast asleep. Mrs Graham later recalled, 'There was a rap at the door. It was a passenger whom we had met shortly after the ship left Southampton and his name was Washington Roebling. He warned us of the danger and told us that it would be best to be prepared for an emergency. We lost no time after that to get out

167

into the saloon. I met an officer of the ship. "What is the matter?", I asked him. "We have only burst two pipes," he said. "Everything is all right, don't worry." ' The Graham ladies remained in the warmth of the saloon.

Back up on the Boat Deck, Second Officer Lightoller had finished uncovering the lifeboats on the port side of the ship and was wondering what to do next. He saw Chief Officer Wilde heading for a door leading below and ran over. Bellowing to make himself heard above the noise of the steam, Lightoller asked if the lifeboats should be swung out ready to be loaded. Wilde shook his head and passed on. Then Lightoller saw Captain Smith and repeated his question. Smith nodded, so Lightoller went back to the port-side lifeboats and began swinging them out, just as the main body of first-class passengers started arriving on the Boat Deck, shepherded up from their cabins by the stewards.

CHAPTER 6

THE FIRST SOS

At about the time that the boats began to be swung out, 12.20 am, Jack Phillips in the radio room got some good news. He received a reply to his repeated distress signals. It came from the SS *Frankfurt*, a German liner. The radio operator on the *Frankfurt* told Phillips that he had received the distress signal, then added 'OK, stand by'. Then silence. Phillips went back to sending the distress signal CQD and the *Titanic*'s position at 41 46° N and 50 14° W.

Everyone had forgotten about Fleet and Lee up in the crow's nest. They had been puzzled when the ship stopped, but not worried. Now that they heard the steam being let off they knew something was wrong. They scrambled down the ladder to the deck, where the duty quartermaster should have been on duty. According to Fleet, 'There was nobody there. Then the quartermaster come down and said we were all wanted up on the bridge. We went up to the Boat Deck. Before we could reach the bridge I seen [sic] them all – the officers – at the boats, getting them ready and putting them out. I helped.'

Other crew members were also appearing on the Boat Deck. Boatswain Haines and the other crew members who had their cabins in the forward area of Deck G had initially gone back to their cabins after Wilde had left them. But with the ship now listing slightly to starboard Haines had become concerned. He decided to go up on deck to see what he could find out. Haines bumped into Thomas Andrews of Harland and Wolff and asked him what was happening. Andrews told him, so Haines raced back down to his fellow crew members.

Lamp Trimmer Hemming recalled: 'We had gone back to our bunks when a joiner came in and said "If I were you, I would turn out, you fellows. She is making water – one-two-three – and the racket court is getting filled up." He had just gone out when the boatswain Haines ran in. "Turn out, you fellows," he says. "We haven't half an hour to live. And that is from Mr Andrews. Keep it to yourselves and let nobody know." ' In other words, don't tell the passengers in case they panic.

Unsurprisingly Haines, Hemming and their fellows

dashed straight up to the Boat Deck. The ladder they used brought them out at the forward end of the Boat Deck on the port side. There they were quickly grabbed by Second Officer Lightoller and put to work. Hemming would not leave Lightoller's side for the rest of the night.

Lightoller then spotted Captain Smith heading towards the bridge. He approached Smith, cupped his hands around Smith's ears and shouted, 'Hadn't we better get the women and children into the boats, sir?'

Smith nodded. 'Yes,' he shouted back, 'put the women and children in and lower away.'

Lightoller walked back to Boat No. 4. He was worried about lowering a fully-loaded boat from the great height of the Boat Deck all the way down to the sea. The potential for passengers unaccustomed to being in a boat to overturn it was too great. He decided to lower the empty Boat No. 4 down to A Deck and then to load women and children into it from the promenade area there. As the boat was being lowered down, Lightoller sent a man down to supervise the loading. The man was soon back. Although that section of A Deck was a promenade, the sides were lined with glass panels to protect walkers from inclement weather. They had been locked shut earlier that evening as the temperature fell. Lightoller packed the man off to find the key and moved on to Boat No. 6.

Before starting work, Lightoller chose a boatswain who was detailed to help him. He ordered the boatswain to take some men and go below to the lower gangway doors, located not far above the waterline. Those doors were to be opened so that women and children from third class could be loaded into the partly empty lifeboats that were being lowered. Lightoller also told the seamen crewing the boats that they were to go to the gangway and 'fill the lifeboats to their utmost capacity'. Unfortunately something went wrong. The gangway doors were never opened and nobody ever again saw the boatswain or his men. Something must have happened to them. The failure of the doors to open meant that several boats were left partly empty, while the loss of the men that the boatswain took with him left Lightoller short-handed in his work of lowering and crewing the boats. He would later grab any crew member he could find.

To the great relief of everyone on board, the steam had now finished escaping from the boilers. The deafening noise stopped abruptly and it was at last possible to communicate with each other.

Captain Smith walked into the first-class saloon at about 12.20 am, where Alfred Omont, the card-playing Frenchman, was still hanging about waiting for news. So were about 60 other first-class passengers. To Omont, Smith appeared quite unconcerned:

The captain and first officer [Murdoch] came down. The captain was chewing a toothpick and he said 'You had better put your life preservers on. Just as a precaution.' Then I went to my cabin and I put on my lifebelt. I went to the Boat Deck, but it was deadly cold, so I came back to my own cabin and put on my overcoat. Then I came up and back to the Boat Deck.

At 12.25 am Phillips received another response to his distress calls. This time it was from the RMS *Carpathia*, a Cunard liner of 13,500 tons steaming from New York to Fiume (then in Italy but now in Croatia and renamed Rijeka). *Carpathia*'s signal said that she was about 60 miles (96 km) to the south-east, but had turned around and was now heading towards the *Titanic*. Phillips scribbled the message down and gave it to Bride to take to the captain. Bride found Captain Smith on the starboard side of the Boat Deck supervising the loading of a lifeboat. He handed the message to Smith, who read it, nodded, and waved Bride back to the radio room.

Also at 12.25 am, Quartermaster George Rowe was summoned from his lookout position at the stern of the ship. He was wanted on the bridge and was instructed to bring with him the Socket distress signals and their launching apparatus. 'I took them to the bridge,' Rowe recalled later, 'and turned them over to the fourth officer [Boxhall]. I assisted that officer to fire them, and was firing the distress signals until about five and twenty minutes past 1.'

These signals were a relatively new invention and were designed to replace the rockets and gunfire stipulated in the Board of Trade regulations as being a recognized call for help at night. The signals were explosive charges that were fired from a mortar. They went up to a height of about 700 ft (213 m) and then exploded with a loud report, similar to that of a gun, and emitted stars identical to those thrown out by a signal rocket. The stars were white, as the Board of Trade stipulated for distress rockets. On a clear night, such as this was, they should have been visible over a distance of up to 20 miles (32 km) and heard up to 5 miles (8 km) away. To be recognized as distress signals, the Socket signals had to be fired off at regular intervals of a few minutes. A single signal would not have been identified as a cry for help, nor would a series of signals at random intervals. Boxhall knew his business. He fired them at steady five-minute intervals.

The last one would be sent off at 1.40 am, after he himself had moved on to other duties.

At 12.30 am, radio operator Phillips decided to try contacting the SS *Frankfurt* again. He sent the sinking ship's position, then added: 'Tell your captain to come to our help. We are on the ice.' There was no reply.

Sometime around 12.30 am, Chief Third Class Steward Kieran returned to where Steward John Hart was waiting patiently while trying to keep his restive third-class passengers and the increasing numbers of new arrivals in check. Hart remembered:

I was told, 'Pass your women and children up to the Boat Deck.' Those that were willing to go up to the Boat Deck were shown the way. Some were not willing to go to the Boat Deck and stayed behind. Some of them that went to the Boat Deck found it rather cold. They saw the boats being lowered away, and thought themselves more secure on the ship, and consequently returned to their cabins. I heard two or

three say that they preferred to remain on the ship
than be tossed about on the water like a cockle shell.

Hart had led his group of women and children up to
C Deck, then along the second-class promenade into a
first-class corridor and up the main first-class staircase
to the Boat Deck. He noted, in passing, that all the doors
and barriers that normally separated the various classes
of accommodation had been removed or propped open.
He saw other third-class stewards also leading women
and children up to the Boat Deck. One, a steward
named Cox, was leading a particularly large group of
third-class passengers through the unfamiliar rooms
and corridors of the first-class area. Having delivered
their charges, Hart, Cox and the others hurried back
down to their respective third-class areas.

Other third-class travellers were trying to make their
own way up to the boats. They did not know the way and
several got lost in the maze of passages and corridors.
One group of Irish passengers came up a crew staircase

and found an entrance into a first-class corridor. Their path was blocked by a member of the crew. He told them that they had to wait below until summoned by their own steward. Three women from Longford were at the front of the group, and despite their protestations, the crewman was firm. Then a big Irishman further down the stairs yelled out, 'For God's sake, man. At least let the girls through.' The crewman did so and the women in the group passed on up towards the Boat Deck.

Among the men encouraging the third-class women to go up to the Boat Deck was the chief baker, Joughin, now rather the worse for the whisky he had been drinking. He was none too gentle with the women, especially those who could not speak English and did not understand what he was saying. He practically dragged several up the stairs and one woman found herself thrown bodily into a boat. 'I myself and four other chaps,' Joughin later recalled, 'went about and found some women sitting, squatting on the deck. They did not want to go. We picked them up and forcibly carried them up to the Boat Deck and some children. We threw them into a boat. Then we looked for more.' Joughin kept at his work for half an hour, by which time he felt in need of some more whisky and went back down to his cabin.

'Sinking. Please come'

It was 12.36 am when Phillips received the next answer to his distress calls. This time it was from SS *Prinz Friedrich Wilhelm*, the prestige liner that was the flagship of the North German Lloyd Line. The German liner gave its own position – she was over 120 miles (193 km) away. Phillips responded. 'Are you coming to us?'

Instead of a reply from the *Prinz Friedrich Wilhelm*, Phillips heard the *Frankfurt* cutting in again, asking: 'What is the matter with you?' Phillips sent back, 'We have collision with iceberg. Sinking. Please tell captain to come.' There was no reply from either German ship. Phillips went back to sending out the distress signal and his position repeatedly.

Then Bride had an idea. An international meeting of radio companies and governments had decided that CQD, the usual distress signal, was too difficult for amateurs to send or recognize easily. They instead suggested that a new distress signal should be introduced: SOS. This was said to stand for Save Our Souls, but more importantly consisted of three short dots, three long dashes, and three short dots. Even an absolute beginner could manage that.

'Why not send SOS?' said Bride. 'It's the new call and this may be your only chance to send it.' Phillips laughed and sent the signal – it was the first time it had been sent in a real emergency. The time was 12.45 am.

At some point in these radio signals, Bride again went out to find the captain and give him an update. However, he found the captain was very busy and appeared to be rather distracted. Bride decided not to bother him and instead went back to the radio room.

On the starboard side First Officer Murdoch and Fifth Officer Pitman were also lowering boats. At 12.45

am, Murdoch put Boat No. 7 into the water. On board were the movie star Dorothy Gibson, her mother and 26 others – though the lifeboats had the capacity to hold 65 adults each. The Frenchman Omont was one of those in the boat. He later felt the need to explain his actions and recounted:

I was on the Boat Deck when a boat began to get down. The 1st Officer [Murdoch] saw me and asked me if I wanted to get in. Some of the passengers shouted to me not to get in as they had such confidence in the ship. I saw that the sea was very calm and on reason I thought it better to jump into the boat and see what would happen. I jumped and landed any how. There were twenty-nine of us in the boat. The boat could not have held more than thirty in any case. I personally consider and state that the idea of putting seventy people in a boat is ridiculous. I have a photograph in my possession which shows how ridiculous.

Omont was both right and wrong in his statement. Others confirmed that the boats went off only partly filled because passengers did not want to get into such apparently flimsy craft on a bitterly cold night. They also verified that Murdoch asked if any men wanted to get in once it became clear that no other women would do so. He is, however, quite wrong about the capacity of Boat No. 7. It was rated as holding 65 adults, but later other boats of identical size carried more than 70 quite easily.

No. 7 was not the only lifeboat to pull away from the *Titanic* significantly less than full. Given that there were nothing like enough lifeboats on board in the first place, it might seem odd that officers as experienced as Murdoch and Lightoller were allowing boats to pull away only half full. In fact it made perfect sense.

In the first place at this time it was not clear that the ship would sink before another ship arrived to effect a rescue, and Lightoller at least did not know the ship was sinking at all. Moreover all the officers and many of the crew will have been aware of ships that sank with lifeboats still hanging on the davits. Those lifeboats of course went down with the ship and were no use to anybody. It was much more important to get the lifeboats safely into the water, equipped with a competent crew of seamen, than it was to fill them up before they were lowered.

Once the lifeboats were in the water they could be put to good use. They might return to the ship to take off more passengers. Or they could wait nearby for passengers to swim towards them and be taken on board. It was accepted emergency procedure to get the boats into the water as soon as possible, and this is what the officers of the *Titanic* were doing.

The man in slippers

Soon after Boat No. 7 got away, Murdoch moved on to Boat No. 5. He put Third Officer Pitman in the boat to lead the crew, ordered Fifth Officer Lowe to supervise the lowering of the boat and then left to attend to other business. At this point a middle-aged man in a heavy overcoat and slippers strode up and began giving orders to the crewmen lowering the boat. According to Lowe's account:

> *He was at the ship's side, he was hanging on to the davit. He was shouting 'Lower away. Lower away. Lower away.' He was interfering with my duties. I shouted at him 'If you will get the hell out of that I shall be able to do something.' He did not make any reply, but looked at me. I said 'You will have me drown the whole lot of them.' Then he walked away. I went on lowering the boat.*

Lowe had not recognized the man in slippers, but one of the first-class stewards who was on the Boat Deck had. It had been Joseph Bruce Ismay, chairman of the White Star Line. For a junior officer to shout and swear at the head of the shipping line was unprecedented and caused a sensation among the crew. What was not clear at the time was that Ismay was behaving in a very strange way. Since he had wandered off the bridge, he had been seen several times either walking aimlessly about, staring out to sea or bustling about in a fit of manic and not entirely productive activity.

Presumably the news that the *Titanic* was sinking had sent him into some form of deep shock. The clash with Lowe was not to be the last, and was very far from being the most controversial thing Ismay would do that night.

A ship is sighted

There was another reason to get the boats off as soon as possible. A rescue ship had apparently arrived. Now the boats could be used to ferry passengers and crew from the *Titanic* to the newcomer as they had been designed to do.

The new arrival had first been seen from the bridge by Fourth Officer Boxhall at around 12.40 am. It was about 4–5 miles (6–8 km) away to the north-east when first spotted. Boxhall remembered,

It was a steamer that was ahead of us. I saw his [sic] masthead lights and I saw his side lights. It [sic] was coming towards us. She got close enough, as I thought, to read our electric Morse signal [the Aldis lamp], and I signalled to her. I told her to come at once, we were sinking. And the captain was standing on the bridge. I told the captain about this ship and he stood with me most of the time when we were signalling. I cannot say that I saw any reply. Some crew on the bridge said that she replied to our rockets and signals, but I did not see them.

Given the height of the bridge from sea level, the horizon on a clear night such as this would have been about 9 miles (14.5 km) away. This means that the mystery ship was about halfway to the horizon when it was first noticed.

Having seen the approaching ship, Captain Smith hurried to the starboard side of the *Titanic* to where Boat No. 7 was resting on the surface of the sea. He

187

pointed towards the approaching lights and, shouting through a megaphone, ordered the boat crew to row to the ship, unload their passengers and come back for more. The French card player, Omont, was in the boat and recorded what happened next.

> When we were in the water we began to row away from the ship. I was rowing. We had about twenty-two women on board. We rowed up to about 150 yd [137 m] from the ship. We saw a light far off, about 10 miles [16 km]. Everyone thought it was another ship – a sailing or a steam boat. We saw it plainly. We all cheered up, thinking we were about to be saved. We saw it gradually disappear. We thought it was a sailing boat that could not move on account of the weather. Later we thought she was an optical illusion.

Omont was once again wrong on a technical point. The light could not have been 10 miles away since from a rowing boat on the sea surface, the horizon is barely 3 miles (4.8 km) away.

Others saw the approaching ship. As he worked to get Boat No. 6 away on the port side, Second Officer Lightoller spotted it:

There was a white light showing about two points on the port bow, whether it was one or two lights I could not say. As to whether it was a mast-head light or a stern light, I could not say. I was perfectly sure it was a light attached to a vessel, whether a steam ship or a sailing ship, I could not say. I could not distinguish any other coloured lights, but merely it was a white light, distinct and plain. It was not over 5 miles [8 km] away. I can recollect seeing it for about half an hour.

Quartermaster George Rowe, who was helping Boxhall send up the distress rockets, also saw the ship, but formed a rather different opinion of it. He later recalled:

I saw the light, just a little on the port bow. It was a white light. I judged it to be a stern light of a sailing ship. I think that there was a ship there. Indeed, I am certain of it, and that she was a sailer.

Able Seaman John Poigndestre, who helped Second Officer Lightoller get the port-side boats away, was absolutely certain of what he saw. He had been at sea for over twenty years in various types of ship and said that he had seen something very similar before. It was, he said, a star. He had previously seen a very similar light in the east, and after a short while it had climbed higher into the sky and shown itself to be a rising star seen through hazy air. This light was seen to the west and was, he thought, a setting star.

Whatever sort of ship the new arrival was, assuming it was not a misidentified star, she did not come any closer to the *Titanic* than about 2 miles (3.2 km). Those on board the *Titanic* who saw the mystery ship and survived were all busy with their own duties. Those whose duty it was to watch the ship did not survive. Nobody is quite certain what became of the mystery ship. Lightoller thought that it had vanished from sight by around 2 am. Boxhall caught a glimpse of it at about 1.50 am when he was in Boat No. 2. She was moving away in a westerly direction. Rowe thinks that he saw it again around 3.30 am, by which time he was in a lifeboat. He said that 'Towards daylight the wind sprung up and she sort of hauled off from us.'

This mystery ship could have saved nearly everyone on board the *Titanic* if only she had come to the rescue. Which ship she was came to be a major focus for investigation after the event, with interesting, if often conflicting, results.

'We will see you through this thing'

One of the first-class passengers who came up on to the Boat Deck at about this time was Major Archibald Butt, a former US Army officer who was working as chief military advisor to US President William Taft. He had been in Europe on holiday and was now returning home. He was spotted by Captain Smith who strode over, pulled Butt aside and told him something. Butt gave a visible start.

A fellow first-class passenger, Irene Harris, told what happened next.

You would have thought he was at a White House reception. A dozen women became hysterical all at once as something connected with a lifeboat went wrong. Major Butt stepped over to them and said: 'Really, you must not act like that. We are all going to see you through this thing.' He helped the sailors rearrange the rope that had gone wrong and lift some of the women in with gallantry. When the time came he was a man to be feared. As one boat with some fifty women was about to be lowered a man, suddenly panic-stricken, ran to the stern of it. Major Butt shot one arm out, caught him by the back of the neck and jerked him backward like a pillow. His head cracked

Joseph Bruce Ismay photographed in 1862. Thirty years later he took over the White Star Line from his father as it came under increasing pressure from American competition, a fact that had great influence on the Titanic

As the commodore of the White Star Line Captain Edward John Smith had the traditional duty of commanding a new ship on her maiden voyage. His dog Ben, a Russian wolfhound, did not accompany him on the Titanic

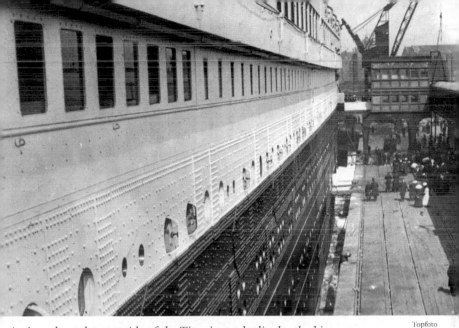

A view along the port side of the Titanic *as she lies berthed in Southampton. The photo was taken from the first-class gangway near the bows, and the second-class gangway can be seen toward the stern*

The Titanic *leaving Southampton. Note the tug visible just behind her. Tugs were always on hand to help liners negotiate the tricky passage down the Solent*

An artwork of the moment of impact. Several witnesses mentioned ice being broken off the iceberg by the strike, while others recalled a succession of impacts as the liner bumped repeatedly against the ice as it passed by

The Russian East Asiatic S.S. Co.   Radio-Telegram.—   526

M16307 ¹

S.S. "Birma".

Words.	Origin.Station.	Time handed in.	Via.	Remarks.
bg to s.	Titanic	11 H.45M.April 14/15 1912.		Distress call Ligs Loud.

Cgd — Sos. from M. G. Y.

We have struck iceberg sinking fast come to our assistance.

Position Lat. 41.46 n. Lon. 50.14. w.

M.G.Y.

The original radio receipt form produced on the SS Birma *as they received the first SOS message sent out by Jack Phillips. This was the first SOS ever to be sent in a real emergency*

M16307 ²

The Russian East Asiatic S.S. Co.   Radio-Telegram.—   527

S.S. "Birma".

Words.	Origin.Station.	Time handed in.	Via.	Remarks.
to	Ship.		SBA	

M.G.Y. We are only 100 miles from you steaming 14 knots be with you by 6-30 our position Lat' 40.48 N. Long. 50.13 w.

S.B.A.

A later distress call from the Titanic *received by the* Birma. *Note that the time is ship's time on* Birma, *not on the* Titanic. *The SOS is supplemented by CQD, the older distress message, and MGY is the code signal for* Titanic

↓ 75 FEET FROM BOAT DECK TO WATER.

A magazine illustration from 1912 showing the launching of the lifeboats. The impression of perfect order shown here was true only of the early stages of the evacuation before the ship began to sink low in the water

A contemporary illustration of the final plunge of the Titanic. The picture gives a good impression of the awful scene, though in a few details it is incorrect. The lifeboats are too large, there were no icebergs in the immediate vicinity and the slight swell shown here did not appear until almost dawn

Survivors on the decks of the Carpathia. *The Cunard liner offered the survivors hot food and drinks as they came on board and provided makeshift sleeping accommodation for them*

Survivors on board the Carpathia. *The passengers on the* Carpathia *held a collection of spare clothing to give to the survivors, several of whom had escaped in little more than their nightwear*

against a rail and he was stunned. 'Sorry,' said Major Butt, 'women will be attended to board first or I'll break every damned bone in your body.'

At about this time Boat No. 1 got away in what would later prove to be controversial circumstances. It was one of the smaller boats, used when in harbour for whatever task required a small boat. It could hold less than half the number of people than a standard lifeboat and, perhaps because of its relatively small and flimsy appearance, was proving difficult to fill.

According to George Symons, the deckhand put in command of the boat, Murdoch was standing on the rail of the *Titanic* shouting out for women and children to come forward when a man came up with two women. The women got in. Two other men approached. One of the three male passengers then asked 'Can we get in?' Murdoch was still scanning the Boat Deck for women and replied rather distractedly, 'Yes, yes, jump in.' One of the three men was the British baronet Sir Cosmo Duff Gordon. The two women in the boat were his wife and maid. Sir Cosmo had a slightly different recollection. According to him there were no other people at all on the deck – not just no other women – and when he had asked if he could get in to the boat, Murdoch had replied 'Oh, certainly do. I'll be very pleased.'

As the *Titanic* began to tilt more obviously towards the bows, Mrs Graham, her daughter and governess were still waiting in the first-class saloon. Then, as Mrs Graham recalled:

Mr Case [Howard Case, an Englishman and another first-class passenger] advised us to get into a boat. 'What are you going to do?' we asked him. 'Oh,' he replied, 'I'll take a chance and stay here.' Just then Mr Roebling came up and told us to hurry and get into a boat. Mr Roebling and Mr Case bustled our party of three into a boat [No. 3] in less time than it takes to tell. The boat was fairly crowded when we three were pushed into it and a few men jumped in at the last moment, but Mr Roebling and Mr Case stood at the rail. They shouted good-bye to us. And what do you suppose Mr Case did then? He just calmly lighted a cigarette and waved us good-bye with it. Mr Roebling stood there too – I can see him now.

Neither Case nor Roebling survived the night.

One point about the loading of boats on the starboard side made by Mrs Graham is worth remarking. She says that men were allowed to get into the boats at the last moment if there were no women or children nearby. Other survivors confirm that First Officer Murdoch, who was supervising the starboard lifeboats, did allow this to happen. On the port side, Second Officer Lightoller was much stricter. He allowed only women and children to get into the boats, along with two or three members of the crew who were to man the oars and sails.

'The crew was perfect'

One couple who benefitted from Murdoch's attitude was Mr and Mrs Bishop of Dowagiac, Michigan. Dick Bishop was aged just 25 and his wife Helen was only 19. They had gone to Britain on a month-long honeymoon and were now returning home. As first-class passengers they were among the first to arrive on the Boat Deck. Mrs Bishop recalled later that 'I was the first woman in the first boat and I was in the boat for four hours. I was in my bed when we were woken. I got up and dressed quickly. There were few people on deck when I got there and there was no panic. The behaviour of the crew was perfect. My husband, thank God is also saved.' Perhaps Murdoch was moved by the youth of the couple and their devotion to each other.

Back on the port side, Second Officer Lightoller was beginning to regret having sent away the quartermaster with quite so many men to open the doors down below. He no longer had as many deckhands as he would have liked and was having to use stewards, engineers and other crewmen for the various tasks. While the more muscular engineers were used to untie knots, swing out boats and keep the growing crowd at a distance, the stewards were sent pushing through the throng of passengers to find women or children and drag them back to the boats. As they passed, these men shouted continuously, 'Any more women and children for the boat?' On Lightoller's orders they went down to A Deck, but do not seem to have gone any lower.

Boat No. 6 was being lowered down when one of the women realized with some alarm that only one seaman was in the boat. She called up to the deck saying that the women could not handle the boat alone. Lightoller scanned the Boat Deck crowd of passengers, stewards and others.

'Are there any seamen here?' he called. There was a short silence, and then a middle-aged passenger stepped forward.

'I am a yachtsman,' he said. 'If you like, I will go.'

Lightoller gestured at the rope from which the lifeboat

bow was suspended. 'If you're sailor enough to get out on that fall, you can go down,' he said. The man sprang, grabbed the rope and nimbly slithered down it to take his place in the lifeboat. He found the sole seaman on board was Quartermaster Hitchens, the man who had been at the helm when the *Titanic* hit the iceberg.

The yachtsman was Major Arthur Peuchen, vice commodore of the Royal Canadian Yacht Club, no less. Peuchen was no mere yachtsman, he had sailed his boat across the Atlantic and was a noted racer. Although an army officer by training and career, he had earned a fortune after discovering a way to extract valuable acetic acid and formaldehyde from the waste wood chippings of which the Canadian timber industry had so much to spare. The trip on the *Titanic* was his fortieth voyage across the ocean on a liner. He was the only male passenger that Lightoller allowed into a boat all evening.

Thomas Andrews, the designer of the *Titanic*, reappeared on deck about this time. He and his team from Harland and Wolff had been hard at work down below keeping the electricity generators and pumps going. Now he had come up to the Boat Deck, apparently to see how things were going and to talk to the captain. Before going below again he walked along the deck urging the women to get into the lifeboats. Many were understandably reluctant to leave their husbands, or to trust themselves to the apparently fragile rowing boats rather than the sturdy and nicely heated liner.

Mary Marvin was one of those wives. She was only 18, the same age as her husband Daniel. 'It's all right girl,' Mr Marvin said as he helped her into Boat No. 10. 'You go and I will stay a while.'

Third-class passenger Adolf Dyker of Connecticut led his wife Anna to Boat No. 16. 'I'll see you later,' were his last words as he pushed her over the gunwale.

Not all wives were convinced by such comments from their husbands. Mrs Tillie Tausig of New York got her teenage daughter Ruth into Boat No. 8, then turned back to her husband Emil. The pair began to argue about whether she ought to go with Ruth or stay with Emil.

Suddenly Mr Tausig caught the eye of somebody behind Mrs Tausig and nodded. Two pairs of muscular arms grabbed Tillie by the shoulders, picked her up and threw her into the boat. The boat was already 10 ft (3 m) down the side of the ship before she could recover herself. Emil gave her a quick wave, then turned around and walked out of sight. None of these three husbands would survive the night.

'They seemed oblivious to what was going on'

At about this time, the retired American army officer and historian Colonel Archibald Gracie went down to B Deck with a steward to help carry blankets up to be put into the boats – these were intended to keep people warm in the freezing night air. On his way back up he passed the first-class smoking room on A Deck. His recollection of the encounter was as follows:

There all alone by themselves, seated around a table, were four men. Three of them were personally well known to me: Major Butt, Clarence Moore and Frank Millet, but the fourth was a stranger whom I therefore cannot identify. All four seemed perfectly oblivious to what was going on decks outside. It is impossible to suppose that they did not know of the collision with an iceberg and that the room they were in had been deserted by all others, who had hastened away. It occurred to me at the time that these men desired to show their entire indifference to the danger and that if I advised them as to how seriously I regarded it, they would laugh at me.

On the port side of the Boat Deck, Second Officer Lightoller was peering down a narrow staircase that was usually for the use of crew only. It reached all the way down to E Deck. He could see water swirling around the base of the stairs. For the first time he realized that the ship was not just in trouble, but it was actually sinking. He began to shout louder for women and children, getting his men to drag, push, cajole and shove to part couples and get the boats filled up before they were lowered. He had accepted by now that the quartermaster he had sent below had been unable to open the lower doors, so passengers could not be put into boats that way. They had to get in from the Boat Deck.

At 1 am Phillips got a reply from *Titanic*'s sister ship the *Olympic*. Phillips replied: 'We are sinking head down. Come soon as possible. Get your boats ready. What is your position?' She was even further away than the *Prinz Friedrich Wilhelm*, about 450 miles (724 km).

Also at about 1 am, Third Class Steward John Hart later remembered, there had been some trouble in his area of E Deck with foreign male passengers. An order had been shouted down from above for him to get another group of women and children together to go up to the Boat Deck. By this time the ship was sloping noticeably down towards the bow and the men were getting restive. Hart went around trying to make himself understood by words and gesticulations that women and children should gather in a group by the wide staircase that led up to the third-class smoking room on C Deck. Several of the men tried to push their way forward on to the staircase, and Hart had difficulty making them give way to the women and children.

After some minutes, Hart had gathered together a group of about 25 women and children. He worried that if he delayed any longer the men might make a rush for it and trample aside the women and children. Although he could see several women still in the corridors, he decided to get his group up to the boats. He led them up by the same route as before. He recalled:

We came along the starboard side of the vessel and I took them to boat No. 15. The boat was then right flush with the rail on the Boat Deck. The people were placed in it. There were twenty-two women and three children. And one man who had a baby in his arms. I was ordered into the boat by Mr Murdoch [First Officer]. It was rather dark on the deck and he said 'What are you?'. I said, 'One of the crew. I have just brought these people up.' He said 'Get into the boat with them.' That is how I came to get in. There were a number of men on the deck. They were standing in absolute quiet. There were some women with their husbands. They would not get in the boat. A cry was raised to ask if there were women and children, the boat was almost full. The boat was lowered. It was about a quarter past one.

Not all third-class foreigners were unable to speak English. August Wennerstrom, from Sweden, knew enough of the language to understand what the stewards were trying to achieve. He rounded up a group of about a dozen Swedish women and convinced them to follow him up to the Boat Deck. He did not know the way, and took a few wrong turns, but eventually got there by going upwards every time he came across stairs. Once on deck, Wennerstrom did his best to push men aside and shout in English to get the women at the front of the growing crowds and so get them into boats. He had soon seen enough to become convinced that the ship was sinking. He decided to go back down to third-class areas to explain this to the large numbers of people below who still did not realize the seriousness of the situation.

CHAPTER 7

THE BAND PLAYS ON

At about the same time that Steward Hart was trying to get the non-English speakers in third class to understand his instructions, Second Officer Lightoller was approached by a grim-faced Chief Officer Wilde. Where, Wilde wanted to know, were the pistols and ammunition?

All liners carried a small number of pistols, and some had rifles as well. These were usually kept locked away in one of the officers' cabins, and in White Star ships they were the responsibility of the first officer. Wilde was asking Second Officer Lightoller where they were as Lightoller had been first officer when the *Titanic* was fitted out, and so it was to him that the guns had been given in Belfast. Murdoch, although by this time responsible, had never asked Lightoller for them. This was hardly surprising. The guns were for use in dire emergency only and most officers went their entire careers without ever using them.

Now, it was clear, they were needed. Lightoller presumed that there had been some trouble verging on panic somewhere among the passengers, though Wilde did not say so. Lightoller led the chief officer to his cabin where he unlocked the cupboard holding the weapons. Wilde pushed one into Lightoller's hands, together with plenty of ammunition, and told him to put it in his pocket. Wilde then left to distribute the weapons to the officers, boatswains and other senior deckhands. Lightoller returned to his task of lowering the boats.

Boat No. 8 was launched by Lightoller at about 1.15 am. One of the women on board was a young American music teacher named Marie Young, who knew Major Butt, President Taft's military adviser. She recalled:

Archie himself put me into the boat, wrapped in blankets and tucked me in as carefully as if we were starting on a motor ride. When he had wrapped me up he stepped upon the gunwale of the boat and, lifting his hat, smiled down at me. 'Goodbye Miss Young,' he said. 'Good luck to you and don't forget

to remember me to the folks back home.' Then he
stepped back and waved his hand to me.

It was the last anyone saw of Major Butt.

About this time, bandmaster Hartley led his musicians out of the lounge on A Deck and up on to the Boat Deck. They now added popular ragtime numbers to their dance tunes. The deck was by now quite obviously sloping down towards the bows. Loose objects such as trolleys or bottles were rolling forwards, not stopping until they hit a door or wall.

At 1.20 am, Jack Phillips decided to take a break and handed the radio over to his junior, Harold Bride. The two men had received no visitors and had no real idea what was going on outside, though they could feel that the deck was now sloping towards the bows, and could hear the sounds of people moving about and boats being lowered. Phillips went out to have a look around and see what he could discover. While he was gone, Bride received a call from the RMS *Baltic*, one of the White

Star Line's Big Four passenger liners. The message read simply 'Coming to your assistance', but gave no position. She was, in fact, 240 miles (386 km) away.

Phillips came back at this point to report that the ship was sinking, was well down by the bows and listing slightly. Moreover, boats with women and children were leaving the ship. Bride decided to vary the distress signals to emphasize how urgent the situation was becoming. 'We are putting the women off in the boats', he sent at 1.27 am. Three minutes later he sent 'We are putting passengers off in small boats'. At 1.35 am he added 'Engine room getting flooded'.

An hour and a half earlier, Assistant Engineer Herbert Harvey had gone to No. 5 Boiler Room with orders to draw the fires, where Fireman Fred Barrett was already at work doing exactly that. It took about another 20 minutes or so to draw the fires, after which water was fed on to the boilers to cool them down gently and avoid a catastrophic implosion. Barrett did not have a watch on him in his working clothes, but he estimated that almost an hour had passed when new orders came down to Harvey. Barrett takes up the story:

> Harvey asked me to lift the manhole plate off. You lift it off to get to the valves. I do not know what valves it is [sic]. Valves to turn on the pumps or something. I did so. [Assistant Second Engineer] Mr Shepherd was walking across in a hurry to do something and fell down the hole and broke his leg. We lifted him up and carried him to the pump room, me and Mr Harvey. It was difficult to see what with all the water that had been thrown on the furnaces when the fires were drawn to cool them down. The room was thick with steam. We stayed there about a quarter of an hour after that, watching the boilers. Then the bulkhead gave way, I think the bunker, and a rush of water came through the room from the forward end. I did not stop to look. I went up the ladder. Mr Harvey told me to go up.

Neither Harvey nor Shepherd were ever seen again.

The incident in No. 5 Boiler Room is important for reasons that Barrett does not explain. The fact that the watertight bulkhead had given way meant that now the sea was flooding into not five compartments, but six. The ship was going to go down even faster.

Significantly, the direction from which Barrett saw the rush of water coming was the coal bunker, where the slow, smouldering fire had been burning among the coals when the ship had left Southampton. Captain Smith had put to sea after Thomas Andrews from Harland and Wolff had assured him that the fire was not in contact with the outer hull. But the fire had been in contact with the watertight bulkhead. It would seem that the fire had somehow weakened the bulkhead, and so it gave way when the pressure of the water reached a critical point.

Barrett made his way up through the ship. The lower decks were utterly deserted. He did not meet anyone until he got to Shelter Deck, or Deck C, where there were some third-class passengers and third-class stewards. But Barrett did not stop. He knew the ship was now sinking fast. He carried on up to the Boat Deck. He emerged into the cold, night air to find that many lifeboats had already been launched, while others were still on the davits. It was about 1.40 am, he thought.

By the time Barrett got on deck, the tilt of the ship was such that there were very few on board who thought that the ship was not sinking. The passengers had no way of knowing how fast she was going, but most of the crew had a pretty good idea. Second Officer Lightoller had a better idea than most. Every time he got the chance he ran to peer down the stairs that led to E Deck. He could see the green waters creeping up the stairs and did a rough calculation as to how long he had left. An hour he thought, but that turned out to be optimistic.

The American historian Colonel Gracie was walking on the starboard side of the ship when he saw

a young woman clinging tightly to a baby in her arms as she approached near the ship's high rail, but unwilling even for a moment to allow anyone else to hold the little one while assisting her to board the lifeboat. As she drew back sorrowfully to the outer edge of the crowd on the deck, I followed and persuaded her to accompany me to the rail again, promising if she would entrust her baby to me I would see that the officer passed it to her after she got aboard. I remember her trepidation as she acceded to my suggestion and the happy expression of relief when the mother was safely seated with the baby restored to her. 'Where is my baby?' was her anxious wail. 'I have your baby,' I cried as it was tenderly handed along. I remember this incident well because of my feeling, when I had the babe in my care, though the interval was short, I wondered how I should manage with it in my arms if the lifeboats got away and I should be plunged into the water with it as the ship sank.

Back in the radio room, Phillips suddenly leapt to his feet and swore loudly at the radio set, taking Bride totally by surprise. The cause of his anger was another message from the German liner the SS *Frankfurt*. It read: 'Are there any ships with you already? What is wrong with you?' Phillips had had enough of the German who did not seem to have understood the seriousness of the situation. He grabbed the signal key and sent back: 'Shut up you fool. Stand by and keep out.' Then he sent a signal to the *Carpathia*: 'Come as quickly as you can, Old Man.'

Guggenheim accepts his fate

James Johnston, a saloon steward in first class, came across the multi-millionaire Benjamin Guggenheim and his secretary, Victor Gigilo, strolling apparently without concern in the first-class lounge at about a quarter to two. Both men were impeccably attired in full evening dress. Guggenheim accosted Johnston and said:

I think that there is grave doubt that the men will get off safely. I am willing to remain and play the man's game, if there are not enough boats for more than the women and children. I won't die here like a beast. I will meet my end as a gentleman. Now, Johnston, tell my wife if it should happen that my secretary and I both go down and you are saved, tell her I played the game straight out and to the end. No woman shall be left aboard this ship because Ben Guggenheim was a coward. Tell her that my last thoughts will be of her and of our girls, but that my duty now is to these unfortunate women and children on this ship. Tell her I will meet whatever fate is in store for me, knowing she will approve of what I do.

Guggenheim then went off to talk to fellow multi-millionaire Colonel John Jacob Astor.

By this time boats were leaving the *Titanic* more frequently as the officers and men got into the swing of things. It was also becoming more obvious to all on board that the ship was in serious trouble. When the first boats had been sent away, both Lightoller and Murdoch had been unable to find enough women and children to fill them. Murdoch had allowed some men to get into his boats – all of whom were either first-class passengers or engineers no longer needed below – as they were the only men on the Boat Deck at the time. Lightoller was stricter, allowing only women and children into his boats alongside the crew members tasked with rowing the boats. Now there was no shortage of passengers wanting to get into the boats.

A few minutes after No. 15 was launched with Hart and his party, No. 11 was prepared for lowering. Once again, White Star chairman Joseph Bruce Ismay was interfering, but this time with more productive results. Murdoch was again allowing men to board the boat if no women wanted to do so. Ismay spotted two women standing to one side, ran over and began dragging them towards the boat. They were the first-class stewardesses, Mrs Gold and Mrs Martin, who had earlier led the ladies travelling alone to the Boat Deck. Mrs Gold tried to shake off Ismay's grip. 'We are only stewardesses,' she said. 'You are women,' replied Ismay firmly. 'Women and children first.' He pushed them into the boat while Murdoch held the men back.

When boat No. 11 reached the water, it had 70 people on board, more than it was designed to hold. The boat almost came to grief soon after reaching the water, when it drifted under the discharge pipes from the pumps that were spurting water out of the hold – in a desperate attempt to keep the ship afloat for a while longer. The crewmen on board the lifeboat were stewards, not deckhands, and were not skilled in handling boats. They managed to get the boat clear in time, but it had been a close thing.

If saved, inform my sister that I am lost

A woman climbing into one of the boats felt a tap on her shoulder. A man she did not know pushed a sheet of paper into her hand. 'It is for my sister,' he said. It was addressed to a Mrs F.J. Adams of Findlay, Ohio. Some weeks later the woman managed to deliver the letter.

Mrs Adams opened the note and promptly collapsed. It was from her brother, a professional gambler and conman who had not spoken to or sent word to any member of his family for years. The note said simply: 'If saved, inform my sister that I am lost. J.H. Rogers.'

One couple that attracted some attention was the elderly Mr and Mrs Straus, the New York millionaires. Their photos often appeared in newspapers and many passengers recognized them. The American historian, Colonel Gracie, knew them in passing and went to help. He offered to lead Mrs Straus to a lifeboat.

[She said] 'No. I will not be separated from my husband, as we have lived, so will we die. Together.' When he too declined the assistance proffered on my earnest solicitation that, because of his age and helplessness, exception should be made and he be allowed to accompany his wife in the boat. 'No,' he said. 'I do not wish any distinction in my favor which is not granted to others.' As near as I can recall them,

these were the words which they addressed to me.
They expressed themselves as fully prepared to die
and calmly sat down in steamer chairs on the glass-
enclosed Deck Λ, prepared to meet their fate.

The Straus couple were later induced to go up to the Boat Deck to allow Mrs Straus's maid, Miss Ellen Bird, to get into a boat. She got away in Boat No. 8. Charles Stengel reported that at this point:

I can never forget Mr and Mrs Straus, who have been
Darby and John in life and were not separated in
death. The sailors tried to wrench her away from her
husband, but she refused to leave his side. Finally the
sailors had to abandon their task. The last I saw was
the pair standing together arm in arm, Mr Straus
bending towards his wife.

Back on the port side, Second Officer Lightoller and Fifth Officer Lowe were having trouble with the passengers as they began to realize that there were not enough lifeboats. As No. 14 was preparing to be lowered Lowe spotted a young man hiding under one of the seats among the women and children. Lowe grabbed him and hauled him out, ordering him to get back on the ship. The man refused. Lowe pulled his pistol, whereupon the man fell to his knees begging. Lowe was having none of it. He told the hideaway to be a man, and managed to push him back to the Titanic's decks.

Lowe followed him and ordered nearby crewmen to push the crowd back. One of these men was a greaser from the engine room named Fred Scott. According to Scott, Lowe held a gun to the miscreant's head and shouted at the crowd: 'If any man jumps into that boat, I will shoot him down like a dog.'

Perhaps seeing the disturbance, Chief Officer Wilde pushed his way to the boat. He put his imposing bulk beside Able Seaman Joseph Scarrott who was at the stern of the boat and turned to face the crowd of passengers.

'All right, Mr Lowe,' said Wilde. 'You can start loading the women and children now.' After about 20 women had boarded the boat, Scarrott later recalled that 'some men tried to rush the boat, foreigners they were because they did not understand the orders which I gave them and by their dress. I had to use a bit of persuasion with the boat's tiller. Five men got into the boat, but we threw them out again.'

Quickly the boat was filled to overflowing with women and children; Lowe was worried it might collapse under the weight. Then Lightoller told Lowe to take command of the boat and gave the word for it to be lowered. Scarrott recalled that

Mr Lowe was talking to another officer [Lightoller] who said to him, 'You go in this boat and I will go in the next.' Mr Lowe asked me how many were in the boat. I told him as far as I could count there were 64 women and four children, one of those being a baby in arms. It was a very small baby from the way the

mother was looking after it. And we had two firemen
and three stewards.

Warning shots

As the boat began to slide down towards the sea, some
of the men on the Boat Deck again moved forward as if
to try to jump into it. In the lowering Boat No. 14 Lowe
drew his pistol:

I thought if one additional person was to jump into
the overcrowded boat that slight jerk might part the
hooks. I thought, well, I will keep an eye open. So as
we were coming down the decks, I saw a lot of foreign
Latin type people all along the ship's rails and they
were all glaring, more or less like wild beasts, ready to
spring. That was why I yelled 'Look out' and shot –
bang – along the ship's side. I fired these shots without
the intention of hurting anybody. I fired horizontally
between the boat and the ship, there was a gap of
about 3 feet [90 cm]. I shot so for them to know that

I was fully armed. That is the reason. Then I put the gun in my pocket.

The boat got away without further incident.

Lightoller did not go in the next boat away, No. 16, but remained on deck to supervise the loading and lowering of the rest of the boats. Perhaps he was simply trying to get Lowe away.

At 1.45 am or thereabouts the bows of the ship dipped under the water. Five minutes later water flooded into the open area of the third-class promenade behind the forecastle. This was a key event. It meant that the water was now above the tops of the watertight bulkheads and could flow freely around the front of the ship, flooding into areas that until then had held pockets of air. As the water poured in, the bows dipped more steeply down. The stern of the ship began to lift up out of the water, exposing the rudder and propellers to the air. The movement was slow, but unstoppable.

The change was noticed by Chief Baker Joughin, still

in his cabin on E Deck nursing his bottle of whisky. When water started creeping under the door of his cabin, he decided it was time to go. He arrived on B Deck to see the grim sight of empty davits, showing him all too clearly that the lifeboats had gone. Knowing that wreck survivors who managed to grab hold of something that floated in the water stood a better chance of survival than those who did not, Joughin began throwing deckchairs overboard. He had heaved about fifty into the sea when he felt in need of another drink. His path back to his bottle was now blocked by seawater, so he went to the pantry attached to the first-class lounge instead.

The final lifeboat

The last lifeboat to be put to sea was Boat No. 4. This should have been the first to get away, but Lightoller had stopped its launching when it reached the level of A Deck and the windows through which passengers were supposed to enter it were found to be locked. Now the windows were open and Lightoller came down to supervise the loading. As before, he was strict in allowing only women and children on board. The ship was listing to port by this time and there was a gap of some 5 ft (1.5 m) between the window and the boat. Lightoller rigged up a bridge using deckchairs and rope.

The indefatigable American historian Colonel Gracie was among the passengers on A Deck at this point. He remembered that:

Second Officer Lightoller was in command. He was standing on the rail of the boat, while we passed women, children and babies in rapid succession without any confusion whatever. Among this number was Mrs Astor, whom I lifted over the four-feet high rail of the ship through the window frame. Her husband held her left arm as we carefully passed her to Lightoller, who seated her in the boat. A dialogue now ensued between Colonel [John Jacob] Astor and the officer. Leaning out over the rail, Astor asked permission of Lightoller to enter the boat to protect *his wife, which, in view of her delicate condition [she was six months pregnant] seems to me to have been a reasonable request. The officer, intent upon his duty and obeying orders and not knowing the millionaire from the rest of us replied, 'No, sir. No men are allowed in these boats until all women are loaded first.' Colonel Astor did not demur, but bore the refusal bravely and resignedly. Colonel Astor moved away and I never saw him again.*

As Astor moved away, he picked up a young boy and pushed him over the bridge of deckchairs to the boat. The boy was Louis Garrett, a third-class Irish immigrant. Boat No. 4 was eventually filled to Lightoller's satisfaction and lowered into the sea. The distance to the sea was now only about 10 ft (3 m), but Deck A had originally been some 60 ft (18 m) above the sea. The ship was going fast as the boat pulled away. It was 1.56 am.

At about this time the English-speaking Swede August Wennerstrom came back to the Boat Deck, along with more than 200 third-class passengers whom he had convinced that danger was imminent and that they should get on to the open deck as soon as possible.

Second Officer Lightoller meanwhile turned his attentions to the four small collapsible boats, each of which could hold 47 people, which were lashed to the roof of the officers' cabins. Boat D was quickly assembled and lifted down to the Boat Deck to be fitted to the davits vacated by Boat No. 2. As he was organizing deckhands in this task, Lightoller saw a group of about 35 engineers come up from below. He got these big, tough men to link arms to form a human barrier across the deck. It was as well that he did for, seeing the boat, several men made a rush forward to it. The engineers held them back, and Lightoller drew his pistol to wave over his head. 'Any more women and children?' he shouted. Again he sent his team of stewards pushing through the crowds of male passengers seeking women and children for the boat.

Among those who heard the calls for women and

children was Colonel Gracie, then on the starboard side of the ship trying to comfort two first-class lady passengers that he knew, Mrs Caroline Brown and Miss Edith Evans. 'I immediately seized each lady by the arm,' recalled Gracie, 'and hurried, with three other ladies following us, towards the port side. But I had not proceeded half way when I was stopped by a line of the crew barring my progress and one of them told me that only women could pass.' Gracie stepped back and the women were allowed through.

'Your need is greater than mine'

A man came pushing through the crowd with two little boys, the elder of whom looked barely 2 years old. He lifted them up and passed them over the shoulders of the engineers to a crew member, who put them into the boat. As Lightoller was finishing off the loading of the boat, Chief Officer Wilde strolled over. 'You'd better go with her, Lightoller,' said Wilde. 'Not damn likely,' replied Lightoller. There were still two more collapsibles waiting to be launched. Instead Lightoller grabbed two crew members at random and shoved them into the boat; they turned out to be stewards totally unaccustomed to handling boats in the sea but they managed to row the boat away from the liner and keep her afloat. There was now space for just one more woman. The two ladies brought by Colonel Gracie were next in line. 'Your need is greater than mine,' said Miss Evans to Mrs Brown. 'You have children who need you, I have none.' Mrs Brown got into the boat, leaving her friend on the deck. As the boat was being lowered down the side, two men jumped into it from A Deck beneath. Miss Evans could have gone in the boat after all.

On the starboard side, Murdoch was having just as much trouble with Boat C. Unlike Lightoller, he had not got enough tough crew to form a human barrier. As he and his men put the collapsible boat together a seething crowd of men surged around them.

'Stand back. Stand back,' shouted Murdoch. 'It is women first.' The men took no notice of him. When the boat was almost full two male passengers jumped into it. Purser McElroy pulled his pistol and fired two shots into the air. Some crewmen and two passengers grabbed the men in the boats by their legs and hauled them out.

One of these passengers was Eugene Daly, a third-class passenger from Westmeath, Ireland, who had joined the ship at Queenstown. He wrote of the incident on deck:

I was with two Westmeath lady passengers and all three of us knelt down on the deck and prayed. We went to the second cabin deck [the roof of the officers' cabins] and the two girls and myself got into a boat. An officer [presumably Murdoch] called on me to go back, but I would not stir. They then got hold of me and pulled me out. An officer pointed a revolver and said if any man tried to get in he would shoot him on the spot. I saw the officer shoot two men dead because they tried to get into the boat.

Things were starting to get out of hand. At this point,

Chief Officer Wilde strode up and his huge physical bulk calmed things down. The crowd edged backwards.

Together Wilde and Murdoch supervised the loading of the boat after it was swung out on the davits. At one point a third-class steward named Albert Pearcey came up, shepherding forward a Lebanese woman named Darwis Touma and her children, 7-year-old George and 9-year-old Maria. Pearcey had been pushing through the crowded decks looking for women and children when he spotted the two children looking terrified down on B Deck. He had tried telling Mrs Touma that she had to get up on to the Boat Deck, but she spoke no English and did not understand his increasingly frantic gesticulations. Finally losing patience and knowing that time was short, Pearcey picked up the two children in his arms and marched off, hoping that Mrs Touma would follow him, as she did. Now Pearcey heaved the two children into the boat, and helped their mother in after them.

Quartermaster Rowe, who had been firing the Socket distress flares with Boxhall, was put in command of Boat

C. There were no deckhands left, so he pulled in three firemen and the steward Albert Pearcey, who looked muscular enough to handle an oar, to act as crew. The boat was designed to hold 37 people; there were 39 in it as it was put down from the davits into the sea.

One of the women in this boat recognized Murdoch and later wrote: 'I saw Mr Murdoch shoot down an Italian. This officer performed heroic work all through. Just after he had given directions about the men making way for women, someone shouted sarcastically "What about yourself, sir?" "I am going back to the ship in a minute or two," replied Mr Murdock [sic].'

Ismay escapes

As the boat was being lowered, Rowe thought that other than himself and the men he chose as rowers all the people in the boat were women and children. He was surprised to learn as they pulled away from the sinking liner that there were two men present – and even more amazed when he recognized one of them as being Mr Joseph

Bruce Ismay, chairman of the White Star Line. Exactly how Ismay – of all people – ended up in this, the last boat to get away, is not entirely clear. Ismay himself later said: 'The boat was there. There was a certain number of men in the boat, and the officer called out asking if there were any more women, and there was no response, and there were no passengers left on the deck. As the boat was in the act of being lowered away, I got into it.'

First Class Steward Edward Brown was standing nearby and he gave a different version of events. After describing the turmoil among the crowd as the boat was put together and how it was quelled by Wilde, Brown says that Murdoch called for women and children, pushing men aside so that they could get through. 'Mr Ismay,' Brown recalled, 'was standing in the boat receiving the women and children when it was hanging over the side on davits. Then the boat went down as it was full then [sic]. I did not see any more of Mr Ismay. Then I saw four or five women on the deck waiting to get into a boat.'

According to Ismay, he was on the deck as the boat was filled and only stepped into it at the last moment when it was obvious that there were no more women about and only very few men. According to Brown, Ismay had been in the boat all along, there was a great crowd of men about and a few women. Mr Ismay's behaviour would later become a focus for great attention.

At a little past 2 am, Captain Smith went to the radio room. He walked in looking tired and said, 'Well, men, you have done your duty. Abandon your cabin. Now it's every man for himself.' Jack Phillips and Harold Bride looked at each other, then Phillips turned back to the message key to continue tapping out the distress signal. Smith said: 'You must look out for yourselves. I release you from your duty.' Then he left.

He slumped, lifeless, to the floor

Bride began gathering up his and Phillips' things. He draped Phillips' coat over his shoulders, then slung his lifebelt over the back of his chair before slipping into the screened-off bunk area to get his own gear. He was returning to Phillips when he saw the door open and a stoker slip in. The stoker picked up Phillips' lifebelt as if to steal it. Bride was furious; the stoker would have had his own lifebelt but must have lost it somewhere. Bride hurled himself at the stoker as Phillips, realizing what was happening, leapt up and joined the fight. Bride hit the stoker repeatedly on the head until he slumped, lifeless, to the floor.

Bride and Phillips got into their coats and lifebelts, then Phillips returned to the radio set. He sent out CQD a few more times, but the electric power was fading. He gave up and the two men left the cabin to emerge into the frosty night air. They climbed up to the Boat Deck. Bride saw Captain Smith walking towards the bridge, then spotted a group of officers and men wrestling with one of the collapsible boats on top of the officers' cabins. He went up to help them.

In charge of the men working on the collapsibles were First Officer Murdoch on the starboard boat and Second

Officer Lightoller on the port boat. Lightoller got his boat free first. He and his men pushed it over the side of the deck to fall the few feet to the sea that now washed over the Boat Deck. Lightoller turned to lead his men over to help Murdoch. The starboard boat was also free and had been assembled, but Murdoch was having trouble getting it into the water. The list to port had got worse and pushing the boat uphill over the lip of the deck was impossible.

Captain Smith walked calmly along the Boat Deck. He raised his megaphone to his lips and bellowed, 'Each man must try to save himself.'

He repeated the message several times, then turned and headed to the bridge. A British second-class passenger, William Mellors, watched him walk past. He thought that the captain was crying.

Just then, at about 2.10 am, the rising waters reached the forward end of the Boat Deck where Captain Smith was standing on the bridge. The great glass dome over the Grand Staircase gave way, allowing the water access from the Boat Deck right the way down to E Deck. The torrent of water became a mighty waterfall that spread rapidly down into the bowels of the ship. Deep within the *Titanic* something gave way, presumably another of the watertight bulkheads. The *Titanic* gave a convulsive lurch forwards and down, causing a great wave of ice-cold green water to surge back along the Boat Deck, engulfing the bridge and breaking over Murdoch and his men with the collapsible.

Titanic was preparing for her final plunge.

CHAPTER 8

THE FINAL PLUNGE

Second Officer Lightoller saw the wave surging towards him along the top of the officers' cabins. Murdoch and the men helping him were swept away. There was nothing more that Lightoller could do to help the passengers, for every boat was now off the *Titanic*. He considered that if he tried to outrun the wave he was only putting off the inevitable moment when he would be pitched in to the sea, so he dived headlong into the approaching wave. The coldness of the water shocked him. At first he began swimming towards a nearby object, but realized that it was the crow's nest and it would soon go under with the rest of the ship.

Some time later, after he had been questioned exhaustively by reporters, politicians and lawyers, Lightoller was asked by yet another curious gentleman, 'What time did you leave the *Titanic*?' Lightoller eyed the man coldly. He replied:

I did not leave it. It left me. I was sucked to the side of the ship against the grating over the blower for the exhaust. There was an explosion. It blew me to the surface again, only to be sucked back again by the water rushing into the ship. This time I landed against the grating over the pipes which furnish a draught for the funnels and stuck there. There was another explosion, and I came to the surface. The ship seemed to be heaving tremendous sighs as she went down. I found myself not many feet from the ship, but on the other side of it. The ship or I had turned around while I was under water. I came up near a collapsible lifeboat [B] and grabbed it. Many men were in the water near me, they had jumped at the last minute.

One man who jumped in a different direction was the American historian Colonel Gracie. He was talking to a friend, Clinch Smith, on the Boat Deck, rather further back than was Lightoller. He saw the wave coming and, as he later recalled:

looking up towards the roof of the officers' house I

saw a man to the right of me and above lying on his stomach on the roof, with his legs dangling over. Clinch Smith jumped to reach this roof, and I promptly followed. The efforts of both of us failed. I was loaded down with my heavy overcoat and Norfolk coat beneath, with clumsy life-preserver over all, which made my jump fall short. As I came down the water struck my right side. I jumped again as if on the crest of a wave on the seashore. This expedient brought the attainment of the object I had in view. I was able to reach the roof and the iron railing along the edge of it, and pulled myself over on top of the officers' house on my stomach near the base of the second funnel. To my utter dismay, a hasty glance to my left and right showed that Smith had not followed and that the wave had completely covered him.

Another man washed overboard by the surge wave was First Class Steward Thomas Whitely. He had been helping Lightoller load women on to the boats and was with

Murdoch trying to free the last collapsible boat when the wave struck. 'I was dragged around the deck and in some way got overboard and clung to an oak dresser.'

Eugene Daly, the Irishman who had been dragged out of collapsible Boat B by First Officer Murdoch, was struck by the surge wave as it petered out. 'Suddenly I was up to my knees in water. Everyone was rushing around and there were no more boats. I then dived overboard and got into a boat.' It was Boat B, where he was reunited with his two lady friends from Westmeath.

A few moments after the surge wave spent itself, the forward funnel fell forwards. Charles Williams and his son Richard, two Americans living in Switzerland, were standing on the Boat Deck beside it as it fell. Richard recalled:

We were stood on the deck watching the lifeboats of the Titanic. The water was nearly up to our waists and the ship was about at her last. Suddenly the great funnel fell. I sprang aside, endeavouring to pull father with me. A moment later the funnel was swept overboard and the body of father went with it. I sprang overboard and swam to a life raft [collapsible Boat A]. There were five men and one woman on the raft. Occasionally we were swept off into the sea, but always managed to crawl back.

The funnel hit the sea with a tremendous splash and threw out a great wave. 'The funnel fell within four feet of me,' Second Officer Lightoller reported, 'and it killed some of the swimmers. I found myself still clinging to the boat, now capsized.' He hauled himself up on to the keel of the upturned boat to get clear of the water. Others had a similar idea and soon Lightoller was not alone. One of the first to join him was Harold Bride, the junior radio operator.

Also finding refuge on the overturned boat was Yorkshireman Algernon Barkworth. He recorded:

I struggled in the water for several hours endeavouring

to hold afloat by grabbing to the sides and end of an overturned lifeboat. Now and again I lost my grip and fell back into the water. I considered my fur overcoat helped to keep me afloat. I had a life preserver over it, under my arms, but it would not have held me up so well out of the water but for the coat. The fur seemed not to get wet through, and retained a certain amount of air that added to buoyancy. I shall never part with it.

Bandmaster Hartley stopped playing as the deck lurched beneath him. It was, he knew, long past the time for cheery dance music. Those left on board can have been in no doubt now that death awaited them. All the lifeboats had gone and no ship had come to the rescue. It was time for something more spiritual. The band began to play a hymn. Harold Bride said that the hymn he heard was 'Autumn'. Vera Dick, in a lifeboat, said the hymn was 'Nearer My God to Thee'. Jacob Gibbons, a steward rowing the overloaded Boat No. 11, recalled that 'The band was playing and the strains of "Nearer my God to Thee" came clearly over the water with a solemnity so awful that words cannot express it.'

'Gentlemen, I bid you farewell'

The two tunes are fairly similar and, in the circumstances, it is understandable that they could be confused. Both would have been suitable for the grim occasion.

'Autumn' has the refrain:

God of mercy and compassion
Look with pity on my pain
Hear a mournful, broken spirit
Prostrate at Thy feet complain
Nothing can uphold my goings
But Thy blessed Self alone.

'Nearer My God to Thee' has the words:

Though like the wanderer, the sun gone down,
Darkness be over me, my rest a stone;
Yet in my dreams I'd be nearer, my God, to Thee,

Nearer, my God, to Thee, nearer to Thee!
E'en though it be a cross that raiseth me;
Still all my song shall be nearer, my God, to Thee.

Hartley was known to like this hymn and had once said that he wanted it to be played at his funeral. Though nobody can ever be certain, it was probably this hymn that the band played.

When the hymn ended, Hartley lowered his violin and said to the band members, 'Gentlemen, I bid you farewell.' Accounts differ as to what happened to the band next. One witness watching from a lifeboat said that the violinist, Hartley, then began playing another slow tune and that some of the band joined him. The last glimpse this witness had of the band was of the rising seawater lapping around their legs. Another survivor said the band dispersed. A third thought that one band member carried on playing alone until he was washed off the deck. In any case, they had more than done their duty.

One man who had a very narrow escape late in the day was Emilio Portaluppi, who was travelling in a second-class cabin near the rear of the ship. He must have been a very sound sleeper for he had somehow managed to sleep through all the commotion thus far. Something now woke him up and he came to only to find his cabin at a crazy angle and his belongings strewn about. Realizing in an instant what was happening, Portaluppi grabbed his lifebelt and opened the door of his cabin. The corridor outside was deserted. He sprinted up to the second-class promenade area on C Deck, vaulted over the rails in to the ice-cold water, and swam to Boat No. 14. The crew hauled him on board and he was saved. It was about 2.10 am, perhaps a little later.

About this time Charles Stengel, who had earlier witnessed Mrs Straus refusing to leave her husband, jumped into the sea from the ship's rail and swam to a lifeboat where he was hauled on board. He had already put his wife Annie into another lifeboat. When the ship went down, she was convinced that her husband was dead and was overjoyed later to find out he was alive.

Also jumping into the water at this time was the greaser, Fred Scott, who had helped Fifth Officer Lowe hold the crowd back from Boat No. 14. He climbed down ropes hanging from an empty davit on the port side, then let go when he was about 10 ft (3 m) from the water. Rising back to the surface, he swam off towards a lifeboat. It was Boat No. 4. After picking up several more men swimming out from the ship, the man in charge, Quartermaster Perkis, decided the boat was full – in fact it was overfull – and gave the order to row away.

At almost that moment a steward wandered into the first-class smoking room. Standing in the room beside a card table was Thomas Andrews, the engineer who had designed the ship. For the past two hours he and his team of Harland and Wolff engineers had been down below, working frantically with the ship's engineering staff to keep the electricity generators going. Now they had abandoned their task as the angle of the ship grew ever steeper. Andrews' lifebelt was slung over the back of a chair.

'Aren't you going to have a try for it, Mr Andrews?' asked the steward. Andrews did not reply. He simply stared at a painting of Plymouth Harbour hanging on the wall.

The distress calls continue

Among those still on board was senior radio operator Jack Phillips. He had been last seen by his colleague Harold Bride walking aft on the Boat Deck at about 2.05 am; he must have gone back to the radio room. Other ships reported hearing CQD and position signals coming from the *Titanic* right up to 2.17 am.

One newspaper report that was widely circulated in the days after the survivors came ashore quoted a 'Quartermaster Moody' saying that he had seen First Officer Murdoch commit suicide at about this time. 'I saw Murdock [sic] die by his own hand,' went the quote from Moody. 'I saw the flash from his gun, heard the crack that followed the flash and saw him plunge over on his face.' The story has been repeated many times and the incident was shown in the 1997 movie. However, there was no quartermaster named Moody, nor anything like it. The only Moody among the crew was the sixth officer and he did not survive, so he could not have talked to any newspaper reporter.

Several other survivors did mention an officer shooting himself, but none said that they had actually seen the incident take place. One said he saw a dead officer lying on the deck, but he could not identify the man. From his description of the body it might have been almost any crew member.

Among those still on the ship as the stern rose up was Olaus Abelseth, a Norwegian farming in the Dakotas who had been on a visit home to see relatives. He later recalled:

> I was standing there, and I asked my brother-in-law if he could swim and he said no. I asked my cousin if he could swim and he said no. So we could see the water coming up, the bow of the ship was going

down, and there was kind of an explosion. We could hear the popping and cracking, and the deck raised up and got so steep that the people could not stand on their feet on the deck. So they fell down and slid on the deck into the water right on the ship.

As he swam away from the ship, Abelseth sighted a lifeboat and swam towards it. It was collapsible Boat A, the boat on which First Officer Murdoch and his men had been working when the *Titanic* dipped forward and the surge wave washed them and Boat A overboard. The boat had taken on a lot of water from the wave and it had over a foot (30 cm) of water washing about its bottom.

One of the engine crew, Thomas Dillon, was on the poop deck. He had been down below drawing the fires and had come up on deck at about 1.15 am. Since then he had been standing around at the rear of the ship awaiting orders that never came. He later said that there were hundreds of men, but no women, standing quietly on the poop deck. 'There was no commotion, no disorder. They were simply waiting for the ship to go down.' He felt the stern lurch about, then sink rapidly. He was thrown into the water. 'I was sucked down about two fathoms,' he recalled. 'Then I shoved myself off the deck and seemed to get lifted up. When I came up again I saw the after part of the ship come up again. She seemed to almost right herself. Then she took one final plunge. I was swimming for about 20 minutes, then I was picked up by a boat that I afterwards found was No. 4 Boat.' Once in the safety of the boat, Dillon passed out and did not come round for more than an hour.

Yorkshireman Algernon Barkworth was still clinging to the sides of the overturned Boat B when he heard a terrible rumbling noise coming from the ship. 'There were three distinct explosions and the ship broke in the centre. The bow settled headlong first, and the stern last. I was looking towards her from the raft.'

First Class Steward Thomas Whitely was still bobbing about on the floating oak dresser when the final plunge took place. 'I wasn't more than sixty feet [18 m] from the

Titanic when she went down. Her big stern rose up in the air and she went down bow first.' He was then joined on the dresser by two other men who had swum from the sinking ship. Together the three men sat shivering on their precarious perch.

The lights go out

Frenchman Paul Chevre was in Boat No. 7 that had been rowing hard to try to get to the mystery ship. He watched the ship go down. 'Suddenly the lights went out. Little by little the *Titanic* settled down. Strange to say, the *Titanic* sank without noise and, contrary to expectations, the suction was very feeble. There was a great backwash and that was all. In the final spasm, the stern of the leviathan stood in the air and then the vessel finally disappeared – completely lost.' Given the distance that he was from the ship, his comments about a lack of noise and suction must be considered dubious.

Rather closer was Able Seaman Scarrott in Boat No. 14 with Fifth Officer Lowe. Scarrott recalled:

The stern of the ship was right up in the air. You could see her propeller right clear, and you could see underneath the keel. And then she seemed to go in a rush. You could hear the breaking up of things in the ship, and then followed four explosions. There was not much wreckage, not so much as you would expect from a big ship like that.

There was considerable discussion at the time, and much since, as to whether the ship broke in two as it sank or went down intact. Witnesses were divided on the subject. Everyone agreed that the ship remained intact up until the moment that the electricity failed and the lights went out. At that point the stern was lifted clear of the water so that the rudder and propellers were visible, together with a length of keel. The angle of the ship was probably somewhere between 20° and 40°. There then came a loud rumbling or tearing sound that lasted many seconds, but not much more than a couple of minutes. That was followed by three or four loud explosions.

Some witnesses thought that the rumbling was caused by the heavy engines tearing loose from their fittings and crashing forward through the ship and the explosions by the last remaining boilers – kept going to power the electricity generators – imploding as they were struck by the cold seawater. Others said that the noises were caused by the ship breaking in two. They reported that the stern fell back into the sea for a few seconds before it reared up again for the final plunge.

One of the survivors who gave the clearest account of the ship breaking up was the greaser, Fred Scott, who had helped Fifth Officer Lowe hold the crowd back from Boat No.14 and was now shivering wet with cold in Boat No.4. 'We had not rowed far,' he said, 'when she began breaking up. She broke just aft of the after funnel and when she broke off, her stern end came down on a level keel, then she reared up and sank.'

All agree that the stern tilted to port and turned slightly after the lights went out. The stern stood then at a very steep angle, steeper than before. It remained almost stationary for a few seconds and then began to sink down almost vertically into the waters.

Colonel Gracie was still on the roof of the officers' cabin when the ship tilted suddenly and he was thrown into the sea. 'Down, down I went,' he later wrote,

It seemed a great distance. There was a very noticeable pressure upon my ears. When under water, I retained a sense of general direction and swam away from the starboard side of the ship, as I knew my life depended on it. I swam with all my strength and I seemed endowed with an extra supply for the occasion. I was incited to desperate effort by the thought of boiling water from the ship's boilers and that I would be scalded to death. Just at the moment I thought that for lack of breath I would have to give in, I seemed to have been provided with a second wind. I prayed that my spirit could go to my loved ones at home and say 'Goodbye until we meet again in heaven'. When my head at last rose above the water, I detected a piece of wreckage like a wooden crate and I eagerly seized it as a nucleus of a raft to be constructed from what flotsam and jetsam I might collect.

Still on board after Gracie had been thrown into the water was the thoroughly inebriated chief baker, Charles Joughin. He was standing on the crazily tilting poop deck, at the very rear of the ship by its flagpole, when the ship reared up suddenly and began to turn to port.

She took a lurch to port and threw everybody into a great heap. Hundreds of them. Then she tilted badly until the ship was tilting upwards. I climbed over the rail and was standing on the side of the ship. I tightened the lifebelt and put my watch into my pocket. I looked at my watch, it was a quarter past two. I was just wondering what to do next when she went. I was about 150 feet [46 m] from the water when she went straight down. She just glided away, there was no great shock or anything. When I reached the water, I stepped off and I do not believe my head went under the water at all. There was no suction at all. I began paddling about, just treading water.

There were now hundreds of people in the ice-cold sea. Several swimmers began to gather around the waterlogged Boat A. All were men, except two. Rhoda Abbott was aged 35 and had jumped from the rear of the Boat Deck as it slipped below the water in the final plunge. She had been disoriented by the cold, then she saw the boat and swam towards it. The men already on board dragged her over the gunwale. Her two teenage sons who had gone into the water at the same time were not so lucky and were never seen again. Mrs Elin Lindell also reached the boat, but she was exhausted. Too tired to get into the boat, she died of cold while clinging to the gunwale. English-speaking Swede August Wennerstrom reached the boat some time later, hauled himself on board and collapsed in exhaustion.

Exactly how many men reached Boat A is not clear.

All of them were soaked to the skin after swimming in the freezing water, and once in the boat were up to their knees in water. Several of the men died in the hours that followed. As each died, the survivors pitched the body overboard in order to lighten the boat and allow it to ride higher in the water.

Other swimmers made for other boats. Third-class passenger Thomas McCormack, of New Jersey, reached one boat only to be hit over the head with an oar by a man who told him in no uncertain terms that he was not welcome. He eventually reached Boat No.15 and was dragged on board. Bernard McCoy of Carrickatane, Ireland, was similarly rebuffed when he reached Boat No.16. After some considerable argument between himself in the water and some people in the boat he was allowed on board.

Charles Williams, a professional racquet player from London, was in Boat No.14. He saw a bearded man swim towards the boat holding a small child clear of the water. The man passed the child in, and then asked if anyone had seen First Officer Murdoch. Told that Murdoch was not in the boat, the bearded man turned and swam off again. Williams thought the man was Captain Smith.

Colonel Gracie was still trying to gather floating wreckage to form a raft. He later said:

What impressed me at the time that my eyes beheld the horrible scene was a thin, light-gray smoky vapor that hung like a pall a few feet above the broad expanse of sea that was covered with a mass of tangled wreckage. That it was a tangible vapor and not a product of imagination, I feel well assured. The agonizing cries of death from over a thousand throats, the wails and groans of the suffering, the shrieks of the terror-stricken and the awful gaspings for breath of those in the last throes of drowning, none will ever forget to our dying day.

After a while he spotted an overturned lifeboat and began paddling towards it. It turned out to be collapsible Boat B with Second Officer Lightoller, the radio operator Bride and others hanging on to it.

Time for the lifeboats to return

Now was the time for the lifeboats to go back and pick up the survivors swimming in the water. This was the reason why Lightoller, Murdoch and other officers had been quite content to send off lifeboats only partly filled. Their priority had been to get all the boats away before the *Titanic* sank so that they would not go down with the ship. With the liner beneath the waves, the task of those in the lifeboats was to go back. They should haul on board anyone they could find and, when the boats were full, allow others to hang on to the ropes slung around the boats for just this purpose. If any of those in the water began to tire or passed out, they were to be pulled on board and one of those in the boat slip into the water to take their place. After other shipwrecks people had survived for hours, days even, in exactly this way.

Fifth Officer Lowe knew his duty. Unfortunately his Boat No.14 was already overloaded. Within sight across the flat, starlit sea were a number of other boats and Lowe could see that they were not full. The boats turned out to be Nos.4, 10 and 12, plus collapsible Boat D. Able Seaman Scarrott recalled what happened next:

Mr Lowe ordered four of the boats to tie together by the painters. He told the men that were in charge of them, the seamen there, what the object was. He said 'If you are tied together and keep together, if there is any passing steamer they will see a large object like that on the water quicker than they would a small one.' During the time that was going on – we intended to make fast ourselves, of course, with the four – we heard cries coming from another direction. Mr Lowe decided to transfer the passengers that we had, so many in each boat, and then make up the full crew. It did not matter whether it was sailors or anything, and make up the full crew and go in the direction of those cries and see if we could save anybody else. The boats were made fast and the passengers were transferred and we went away and went among the wreckage. When we got to where the cries were we were among hundreds, I should say, of dead bodies floating in lifebelts. It was very dark and the wreckage and bodies seemed to be all hanging in one cluster.

When we got up to it we got one man, and we got him in the stern of the boat – a passenger it was, and he died shortly after we got him in the boat. One of the stewards that was in the boat tried means to restore life to the man. He loosed him and worked his limbs about and rubbed him. But it was no avail at all because the man never recovered after we got him into the boat. We got two others then as we pushed our way towards the wreckage.

As we got towards the centre we saw one man there. I have since found out he was a storekeeper. He was on top of a staircase, I think, it seemed to be a large piece of wreckage anyhow which had come from some part of the ship. It was wood, it looked like a staircase. He was kneeling there as if he was praying and at the same time was calling for help. When we saw him we were about 40 feet [12 m] from him and the wreckage was that thick. And I am sorry to say there were more bodies than there was wreckage. It took us a good half hour to get that distance to that man. We could not row the boat, we had to push them out of the way and force our boat up to this man. But we could not get close enough to get him right off only just within the reach of an oar. We put an oar on the forepart of the boat and he got hold of it and he managed to hold on and we got him into the boat. Those three survived.

Then we went right around the wreckage and we had a real good look round and I am quite satisfied that there was nobody left. Then we made sail and sailed back to take our other boats in tow that could not manage themselves at all as we had the men in our boat. We was just getting clear of the wreckage.

Fatal decisions

Although Scarrott did not say so, there had been a fatal delay in getting to the rescue. After some considerable time spent shifting the passengers from one boat to the others, Lowe had ordered the burly men in his new crew to rest on their oars for a while. He knew that he could get only fifty or so people into his lifeboat, and that many times that number were in the water. He calculated that if he rowed straight back the boat would be grabbed all at once by hundreds of hands, and that if it overturned or were swamped then nobody would be saved. He decided to wait until the volume of shouting and screaming died down, indicating that fewer people were left alive. Tragically, he miscalculated how long it would take to row the heavy boat the 400 yd (366 m) or so back to the swimmers and more had died by the time he got there than he had anticipated. It had been a heartbreaking decision and one that was to haunt Lowe for the rest of his life. But it had been an understandable one, and one for which nobody ever blamed him.

But there was another feature of the rescue attempt that Scarrott does not mention that did cause outrage when Lowe talked about it later. It occurred as the survivors from Boat No.14 were being transferred to the other boats. As one of the 'women' was clambering out of No.14, Lowe thought that there was something odd about her. He pulled back her shawl to find that she was, in fact, a man. Lowe was so angry that he picked the man up by the belt and hurled him bodily into the boat where he fell heavily. 'He was not worthy of being handled better,' remarked Lowe afterwards.

As Boat No.14 sailed back towards the other boats lashed together, Lowe spotted a collapsible boat that, he said 'looked rather sorry'. It was, in fact, Boat D, which had very few men in it. Those men who were present had no idea how to row and were splashing about badly. Lowe took the boat in tow. It was now getting light enough to see a short distance, and Lowe saw enough to realize that his work was not yet done. He later recalled:

I noticed that there was another collapsible, in a worse plight than this one that I had in tow. I was just thinking and wondering whether it would be better for me to cut this one adrift and let her go, and for me to travel faster to the sinking one, but I thought no, I think I can make it. So I cracked on a bit and I got down there just in time and took off about twenty men and one lady out of this sinking collapsible. I left three bodies in it. I may have been a bit hard hearted, I cannot say, but I thought to myself that I am not here to worry about bodies, I am here to save life and not to bother about bodies. So I left them.

Lowe had more than done his duty. Others were not so punctilious. Third Officer Pitman was in charge of Boat No.5 with about 38 women and six men. His boat could hold at least another twenty people. As soon as the ship had vanished, Pitman said, 'Now men, we must pull towards the wreck. We may be able to pick up a few more.' The crewmen in the boat shipped the oars and began rowing. The women passengers, however, began shouting and arguing. They declared that it was a mad idea to go back as the boat would be swamped. It would, they declared, be better to save those in the boat than try to save more and lose the boat entirely. A terrific argument ensued, which Pitman lost. After making some progress towards the dying swimmers, the rowers gave up.

For more than an hour Pitman sat listening to people moaning and shouting for help. He was distraught. After the rescue, Pitman almost broke down when he was questioned about the inaction of his boat and in later years refused ever to talk about it.

In Boat No.6 things were the other way around. It was the passengers who wanted to go to help and the crew who did not. Quartermaster Hitchens was worried, like Lowe, about the desperate swimmers overturning the boat. He refused to go back. Even when the cries had subsided, indicating that fewer people were still alive, he refused to steer the boat of which he was in command

271

back to the scene of the sinking. Hitchens pointed out that his last orders from Second Officer Lightoller had been to make for the lights of the mysterious ship that seemed to be just a couple of miles away. The Canadian yachtsman Major Peuchens now spoke up, saying that they could no longer see this elusive other ship and so should go back to collect survivors from the water.

A bitter argument raged in Boat No.6 that lasted for some time. In the end both Hitchens and Peuchen had shouted themselves hoarse, and only a first-class passenger named Margaret Brown seemed to have the energy to continue. She spotted another lifeboat; it turned out to be Boat No.16, and instructed Peuchen and a woman to row the boat in that direction. Coming up to No.16, Mrs Brown stood up, pointed at a stoker by the name of Pelham, and told him to come over into Boat No.6 to help the tired Canadian major to row. Hitchens stood up to remonstrate, whereupon Mrs Brown shouted him down, saying that if he did not keep quiet, she would throw him overboard herself.

Another lady, Mrs Lelia Meyer, then accused Hitchens of taking her blanket and drinking whisky. Hitchens swore at her, whereupon the newly arrived stoker growled out, 'Here, don't you know you are talking to a lady?' That seemed to end the bickering in Boat No.6. In company with No.16 it rowed off to see if the mystery ship could be found. Mrs Margaret Brown was later lauded in the press for her actions and dubbed 'The Unsinkable Molly Brown'.

Fourth Officer Boxhall was in Boat No.2, one of the last to get away and already overfull. There was nothing he could do except endure the sounds, knowing he could not help. After a while a dead body drifted into view face down in the water.

But the events in Boat No.1 were far more controversial than any of these. This boat had a capacity of 40 people, but only 12 were on board. This was the boat that had been among the first to get away and into which First Officer Murdoch had allowed Sir Cosmo Duff Gordon to get along with his wife and her maid. Charles

Hendrickson, a muscular fireman that Murdoch had put in the boat to help row it, said 'It is up to us to go back and pick up anyone in the water.' Nobody said or did anything. Hendrickson tried again, talking pointedly at Symons, the deckhand in charge of the boat. It was Sir Cosmo Duff Gordon who replied. He said the idea was ridiculous, he declared it would be too dangerous as the people would swamp the boat. He appealed to Symons, who nodded silently and did nothing.

Meanwhile, the upturned collapsible Boat B was enduring a saga all of its own. The boat had been flipped over when the front funnel collapsed at about 2.10 am. It remained afloat due to the air trapped underneath it, and breathing that air in rather dazed fashion was the junior radio operator Harold Bride. After a while he heard the sounds of people moving about on top of the boat and decided to join them. He ducked under the gunwale to scramble up the side of the boat and squat on the keel.

On the boat Bride found Second Officer Lightoller taking command of a dozen or so men. Also there was the former American army officer Archibald Gracie and the Yorkshireman Barkworth in his fur coat. Other men came swimming to the boat over the following minutes and eventually about forty or so men were perched precariously on the slippery timbers. More and more men, but no women, came swimming over to the boat. Lightoller allowed some on board, but told others they had to stay in the water and hang on to avoid overturning the boat. As men died or fell off, others were allowed on to take their place. Among those who died on the overturned boat was the senior radio operator, Jack Phillips. It was a terrible time. Finally, new arrivals ceased. There was, they all thought, nobody left alive in the water to swim over.

The drunken baker reappears

A great while later, perhaps about 3.20 am, the *Titanic*'s sozzled chief baker, Charles Joughin, came paddling up out of the dark. He had been in the freezing water for over an hour. At first nobody would help him clamber up the slippery, smooth sides of the boat. Then Joughin spotted a kitchen colleague, an entrée chef named John Maynard, who grabbed his outstretched hand and hauled him half on to the boat.

At about 3.30 am, Lightoller realized that Bride was on the boat. He asked if they had managed to get in radio contact with anyone. Bride told him that several ships were on their way, but that the nearest one, *Carpathia*, would not arrive before dawn. Someone suggested they should pray. They recited the Lord's Prayer.

Not long afterwards a slight breeze got up, and small waves began to wash over the top of the overturned boat. Lightoller realized that if the men did nothing it was only a matter of time before the air trapped underneath leaked out and they were all thrown into the freezing water. 'We must all stand up,' called Lightoller. There was little enthusiasm among the cold, wet and exhausted men. Then the drunken Joughin piped up: 'Yesh. We musht all obey the officer.' Reluctantly the men stood up. Lightoller arranged them in a double row, one each side of the keel and all facing forwards. There were, by this time, only 30 left alive. As each wave approached, Lightoller would shout out 'lean left' or 'lean right' as the occasion demanded. The boat stayed afloat.

Sometime before dawn, Symons, in charge of Boat No.1, was finally persuaded by the fireman, Hendrickson, to go back. By this time there was no calling or shouting of any kind. When the boat reached the edge of the mass debris, all that could be seen were dead bodies and floating wreckage.

At this point one of the firemen, Robert Pusey, turned to Sir Cosmo Duff Gordon and asked, 'I suppose you have lost everything?'

'Of course,' replied the wealthy baronet.

'But you can get more,' said the fireman. 'We have lost our kit, and the company won't give us more.'

'Very well,' replied Sir Cosmo, 'I will give you a fiver each to start a new kit.' Five pounds was not much to a wealthy man such as Sir Cosmo, but it was a sum that would come back to haunt him with a vengeance.

At about 3.45 am a flash was seen on the horizon to the south-east. Some thought it was a falling star, others a flash of lightning. The men standing on the overturned Boat B were too busy keeping upright to notice it at all. In fact it was a rocket sent up by a steamer going hell for leather. A genuine rescue ship was at last arriving.

CHAPTER 9

THE RESCUE

While the tragic events were unfolding on the *Titanic*, the radio signals sent out by Jack Phillips and Harold Bride were pulsing through the air. The distress calls were clear and unambiguous. Help was needed, and it was needed quickly.

There was no law that said that anybody receiving the signal had to go to the aid of the *Titanic*, but the customs of the sea demanded that they did. All seamen knew these customs of the sea and abided by them. The customs are quite distinct from the Maritime Laws that have been enacted by most nations with a coastline and which regulate – often in tortuous detail – how ships should be run, manned, equipped and operated. The customs of the sea are, by contrast, simple and straightforward. They developed over thousands of years, passed down from one generation of seamen to the next as shipping developed from the oared galleys of the Phoenicians through the sailing merchant men of the Roman Empire and so through the Mediaeval period and into the modern era.

The customs arose to make life at sea as safe and secure as possible, while laying great responsibilities on those who went down to the sea in ships. There were four customs that were generally held to apply in times of disaster. The first was that the captain had the duty of being the last to leave a sinking ship. He was expected to ensure that everyone who could be saved was saved before he himself left. From all the evidence available, it would seem that Captain Smith performed this duty as the *Titanic* foundered. A second custom observed on the *Titanic* was that of sending off women and children first. A third that was fortunately not called for was that if survivors were in an open boat for any amount of time and ran short of food they would draw lots to see which one should be killed so that the others could eat their flesh, drink their blood and so survive. The fourth of the customs relating to disaster at sea was that any nearby ship that could render assistance should do so regardless of the cost in time, money or expense to itself or its owners.

It was with this fourth custom in mind that Captain Smith ordered Phillips and Bride to send out repeated demands for emergency assistance. Similarly, the Socket signals were fired into the air to attract the attention of any ship without radio – and most smaller ships lacked such equipment back then – that might be close enough to see or hear them.

Radio equipment in 1912 was in its infancy and was not always reliable. It was possible for signals to travel hundreds of miles, or to fade away after only a few dozen. Moreover, most ships carried radio operators who were employed not by the shipping line but by the radio company. Those on the *Titanic* were employed by the Marconi company. There was no set routine as to when the radio men should be on duty, with captains and radio men on each ship sorting out what suited them best.

On larger liners such as the *Titanic*, it was usual for there to be two radio men aboard. They took it in turns to be on duty, though that did not mean that the radio was on the whole time. It might be switched off for maintenance or if the man on duty was called away by the captain. Sometimes the term 'on duty' was interpreted as being asleep in a makeshift bed in the radio cabin in case somebody wanted a message sent urgently.

Smaller liners and most freighters had only one radio man on board. He was expected to be available to send a message at any time that the captain ordered him to do so, but otherwise his working hours were usually left up to him. On liners, radio operators often worked from mid-morning to mid-evening as those were the times that passengers tended to want to send messages that would be delivered as telegrams to friends, family and business contacts.

Radio men on ships tended to lead essentially solitary lives. Almost invariably they were young, unmarried men who had been attracted by the excitement of what was then a new technology and who did not have families to keep them on shore. They were often in the habit of chatting to each other over the air waves in the evening when atmospheric conditions were at their best for clear, long-distance radio transmissions. Most such chatter would end by the late evening as the radio operators needed to be up and on duty by mid-morning.

The radio equipment carried by the different ships varied widely. Some were powerful enough to send a signal up to 1,000 miles (1,600 km); others were rated to as little as 100 miles (160 km). They also varied considerably in reliability and breakdowns were frequent.

In smaller ships, the radio operator slept in the radio cabin. Some men switched their equipment off when they went to bed so that they could sleep undisturbed; others left the equipment on so that any strong incoming signal would wake them up. Older ships, which had added radio equipment long after they were built, had a wooden structure erected somewhere on deck where the radio man lived and where he worked the equipment. These rooms were termed 'shacks' and many radio men called their working rooms 'radio shacks' even if they were regular cabins.

The shore stations that received the messages from ships for onward transmission to land addresses as telegrams, or sent messages to the ships, were manned 24 hours a day. The shore station that served the area where the *Titanic* went down was Cape Race, on the most westerly point of Newfoundland, then a self-governing dominion within the British Empire and not part of Canada. It was linked by extensive telegraph cables to most of the large cities in North America. It was to Cape Race that Phillips was sending passengers' private messages when the *Titanic* hit the iceberg.

Distress calls received

Phillips ceased sending the passenger messages as soon as the captain warned him to get ready to send a distress signal if required. He sent his first CQD signal at 12.15 am, *Titanic* time, and although he received no reply it was in fact picked up by two different ships. The first was *La Provence*, a small French liner steaming east towards Marseilles in the Mediterranean. She was about 1,000 miles (1,600 km) east of the position given by the *Titanic*. The captain believed he was too far away to give assistance, but sent the signal out as a repeat to alert other ships that may not have heard the original signal.

The second ship to pick up *Titanic*'s initial distress call was the SS *Mount Temple*, of 9,000 tons, belonging to the Canadian Pacific Line heading west to Montreal. She was about 55 miles (88 km) to the west-south-west of the *Titanic*. The radio operator on the *Mount Temple* immediately sent back a reply stating that he had heard the distress call and was off to tell his captain. The *Titanic* never received this message. True to his word, however, the radio man did go to the bridge at once. The captain, John Moore, was not there so the radio man hurried down to his cabin where he shook Moore awake and shoved the message from *Titanic* into his hand. Moore switched on his light, read the message and sprang into instant action.

He first grabbed the speaking tube that led to the bridge and ordered the second officer, who was officer of the watch, to put about and steer east. He then sent the radio man back to listen for more signals and roused the first officer, telling him to get the charts out that covered the *Titanic*'s position. Moore later recalled:

> *After I was sufficiently dressed I went down to the chief engineer and I told him that the Titanic was sending out messages for help. I said 'go down and try to shake up the fireman and, if necessary, give him a tot of rum if you think he can do any more'. I believe this was carried out. I also had him inform the fireman that we wanted to get back as fast as we possibly could.*

Moore estimates that by 12.45 am he was steaming at his ship's top speed of 11.5 knots on a course directly towards the *Titanic*'s reported position. At that speed it would take more than four hours to reach the *Titanic*. A lookout in the crow's nest was under orders to watch for distress rockets in the sky and for ice in the water.

Moore himself was on the bridge for the rest of the night, supervising operations. He had by this time roused all his officers and deckhands. They were given orders that lifeboats were to be uncovered and swung out ready to be lowered at a moment's notice. Ladders were prepared for lowering so that survivors from lifeboats could clamber up them to safety.

Although the *Mount Temple* was primarily a freighter, she did have accommodation for several hundred steerage passengers and stewards to service them. The stewards were woken up and told to get ready to receive survivors. Piles of blankets were readied, empty cabins opened up and the beds made. The cooks began preparing food and water. Other stewards were posted in the corridors to keep the existing passengers in their cabins and below decks. In the event, none of the *Mount Temple*'s passengers even woke up as their ship turned about and began ploughing back the way they had come.

Next to pick up the distress signal at 12.18 am was the SS *Ypiranga*, a German freighter of 8,100 tons belonging to the Hamburg-Amerika Line. Her radio operator could not hear the position being sent by Phillips and so the ship's captain could not work out which way to go to offer assistance.

Seven minutes later at 12.25 am the Cunard liner RMS *Carpathia* came on the air. The ship was a small 9,000-ton liner heading from New York to various Mediterranean ports, ending at Fiume (now Rijeka). She was 58 miles (93 km) south-east of the *Titanic*'s reported position and heading east. The liner's radio operator, Harold Cottam, had some time earlier picked up messages being sent from the shore station at Cape Cod to *Titanic*, but *Titanic* had not answered and Cottam had supposed, correctly, that the signals had not been received. He was now ready for bed but, before turning in, thought he would try to send the signals on to the *Titanic*. He called the *Titanic*, then sent, 'I say Old Man do you know there is a batch of messages coming through for you from Cape Cod?'

He was startled when Phillips sent back: 'Come at once. We have struck an iceberg. It is CQD Old Man. Position 41°46'N 50°14'W.'

'Shall I tell my captain?' signalled back Cottam. 'Do you require assistance?'

The reply was swift. 'Yes. Come quick.'

Cottam ran to the bridge of the *Carpathia* where he found First Officer H. Dean. Even before Cottam had finished speaking Dean was bundling him down the passageway to the captain's cabin. *Carpathia*'s commander, Captain Arthur Rostron, had gained some minor fame in 1907 when he sighted and sketched a large and unknown creature that he identified as the sea serpent of maritime legend. He had been at sea since 1886 and was a highly experienced sea officer.

The *Carpathia* reacts

Captain Rostron told Dean to get the *Carpathia*'s position and then to come to the chart room. Pulling on his coat, Rostron grabbed a passing deckhand and told him to rouse the boatswain and get all the lifeboats uncovered and swung out. The man goggled at Rostron in surprise and fear, but the captain quickly explained that the *Carpathia* was in no danger and the man ran off to do as ordered. Having pulled on some clothes, Rostron rushed to the chart room, sending off another man to wake the chief engineer. He had just worked out a course and given it to the first officer, when, as he recalled:

The chief engineer came up. I told him to call another watch of stokers and make all possible speed to the Titanic as she was in trouble. He ran immediately down and told me my orders would be carried out at once. After that I gave the first officer, who was in charge of the bridge, orders to knock off all work which the men were doing – the men on watch – and to prepare all our spare gear and boats and to have them ready. I sent for the heads of the different

> *departments – the head doctor, the purser and chief*
> *steward – I gave them my orders.*

The head doctor was told to get all his medical staff together and prepare to look after any wounded and the many that were expected to be suffering from hypothermia. The purser was to organize his own men to hand out blankets to survivors as they came on board, and to record their names and home addresses. The purser was also to rouse the kitchen and dining staff to get a meal ready for the survivors. The chief steward was to lend as many of his staff to the purser as were needed, and to use the remainder to clear all the saloons, dining rooms, lounges and other public rooms for the reception of survivors. He was also to keep the *Carpathia*'s own passengers in their cabins and out of the way. The first officer was to get the deckhands up on deck to rig up the derricks to have chairs and canvas slings ready for any elderly or injured who could not climb up the ladder to the ship. But first of all, Rostron insisted, the kitchens had to make huge amounts of fresh, strong coffee and give a mugful to each member of the crew.

The German liner SS *Frankfurt*, which was to so annoy Jack Phillips on the *Titanic* with its odd questions, first heard the distress signal at 12.30 am. She was more than 140 miles (225 km) away to the south-west heading for Boston from Hamburg, being a liner of the North German Lloyd Line. There were clearly some language problems that night, but the captain of the ship knew the customs of the sea as well as anyone. He put his ship about, put extra men to the boilers to increase speed and headed north-east towards the reported position of the *Titanic*.

If the German ship was having language problems, these were trivial compared to the situation on board the Russian ship SS *Birma* of the East Asiatic Line. Captain Leonid Stulpin later wrote down a report that was translated into English by a Russian clerk. The ship was heading from the USA via Rotterdam and Hamburg to Libau (now Liepaja) in Latvia when at 12.45 am it picked up the CQD distress call and the co-ordinates for the *Titanic*. The Russians could not understand the rest of the messages being sent out by the *Titanic* or the answers given, only the CQD. Stulpin recorded: 'We at once altered course for the locality mentioned, increasing the number of stokers so as to come to the aid of the shipwrecked people as soon as possible. Our distance was 100 marine miles.'

It was not language that was a problem for the White Star Liner RMS *Baltic*, but distance. As one of the White Star Line's Big Four, she was a large liner of 23,000 tons with plenty of room to take on board all the passengers and crew of the *Titanic*. She was, however, 200 miles (320 km) to the east. She picked up the distress call at about 12.40 am and at once began piling on extra speed, calling out extra stokers to increase her pace. Her captain estimated an arrival time of about 12 noon.

Other ships were even further away than the *Baltic*, but nonetheless altered course and piled on steam. The customs of the sea could not be denied. Among these were the *Prinz Frederick Wilhelm*, the *Caronia* and the *Olympic*.

A scene of tragicomedy status was played out aboard the small liner SS *Virginian*. She was over 170 miles (273 km) away when the radio operator picked up the distress call. He at once ran up to the bridge where the first officer was on watch. The officer read the message, then glared at the radio man. The *Titanic* was a famous ship and was known to be on its prestigious maiden voyage. He believed that the young radio man was having a joke and ordered him off the bridge. The radio man refused to budge and found himself seized by the arms by two deckhands who lifted him off the ground and at the officer's orders were carrying him towards the stairs that led down to the radio room. The radio man broke free, ran to the captain's cabin and began punching and kicking at the door. Only then did the first officer think that the signal might be genuine. When the captain emerged, groggy with sleep, he gave orders for the ship to head to the position shown at once. The abashed first officer went to work piling on speed.

Not all ships were hurrying to the rescue. Several, such as the SS *Rappahannock* that had passed the *Titanic* shortly before the iceberg collision, did not have a radio. By the time Fourth Officer Boxhall was sending up the Socket distress rockets she was well over the horizon and out of sight. Nor was she alone. The ships *Hudson* and SS *Baron Ardrossan* also lacked radios and had no idea anything was wrong though they were close to the scene of the tragedy.

There was one ship that did have a radio and was close to the scene of the disaster. This was the *Californian*, which had sent ice warnings to the *Titanic* on the afternoon before she hit the iceberg. The *Californian* had been on a parallel course to the *Titanic*, but further north. After nightfall she had come up to the great ice field and been brought to a stop by her commander, Captain Stanley Lord. Up until 11.30 pm the *Californian*'s radio operator had been monitoring radio traffic, but there was nothing urgent or important to be heard. He switched off his radio and went to bed.

Meanwhile, the officer of the watch, Third Officer Groves, had been watching a steamer arriving from the east some miles to the south of the *Californian*. The horizon as seen from the ship's bridge was about 9 miles (15 km) away. The new arrival was closer than that, but not much. As the new ship came steadily on, Groves pointed her out to Captain Lord. Groves judged her to be a liner, but Lord thought she was a freighter. They speculated on her identity.

At 11.50 pm, or thereabouts, Groves saw the approaching ship come to a halt. She was some 5 miles (8 km) away to the south and, as near as Groves could judge, very close to the ice field. He assumed that, like the *Californian*, she had spotted the ice field and stopped while her officers decided what to do. Most of the lights on the ship then went out. As Second Officer Herbert Stone came on watch at midnight, Lord pointed the steamer out to him as part of the usual change-of-watch chatter. Lord then went to the chartroom saying he would sleep on the sofa and that he was to be called if needed, again usual instructions. Left to himself, Stone tried to signal the other ship with an Aldis lamp, but got no response. He was not surprised as she looked to be about 5 miles (8 km) away or maybe more and so at the limit of the signalling light.

At about 12.45 am, Stone saw a rocket go up over the other ship. Three more rockets followed in the next half hour or so. They were low down in the sky, close to the horizon and by the time of the last one Stone had decided that they were coming from beyond the other ship and, indeed, from beyond the horizon. He took them to be the night signals of some other ship, perhaps signalling to the ship that he could see. Nevertheless, Stone decided to alert Captain Lord. Lord asked if Stone thought the fireworks were company night signals; Stone replied that he was not sure. Lord told him to keep an eye on things and to try the Aldis lamp again. Stone did so, but received no reply.

Carpathia reaches unexpected speeds

At 1.25 am Cottam, on the *Carpathia*, heard the *Titanic* signal: 'We are putting the women off in the boats.' When he read this, Captain Rostron decided that the *Carpathia*'s top speed of 14 knots was not enough. He summoned the chief engineer and held a hurried conference. As a result, all surplus systems were closed down. The most significant of these were the hot water systems and central heating that kept the passenger cabins warm. The heat saved was diverted to the engines, squeezing an extra knot out of the old ship. By 2.30 am the extra men shovelling coal had got the *Carpathia* up to 15.5 knots. Never before or again did she reach this speed.

Before then, passengers had begun to emerge from their cabins on the *Carpathia*. Some had been awoken by the cold, others by the noise of stewards moving about. One man whose cabin was just beneath the Boat Deck had been understandably alarmed when he heard the lifeboat above his head being swung out. Several passengers opened their cabin doors, but they were quickly confronted by stewards who told them to get back inside. Those who persisted were told that another ship was in trouble and that the *Carpathia* was going to help. A few evaded the stewards and got up on deck. They found the ship tearing along faster than they had ever known and what looked like the entire crew awake and bustling about at various jobs.

Back on the *Californian*, the officer of the watch, Stone, had been watching the ship to the south of him. At about 2 am he noticed that it was moving. All he could see now was a single bright white light, which he took to be the ship's stern light which seemed to be receding into the distance. He thought perhaps the ship was heading south along the edge of the ice field, looking for a way through. By about 2.30 am the ship had gone out of sight, apparently over the horizon. Stone whistled down the speaking tube to Lord, who mumbled something in reply. Lord later said he had been asleep and remembered nothing of this call.

Captain Rostron of the *Carpathia* later recalled that at 2.50 am,

> *I made out an iceberg a point on the port bow, which I had to steer to keep well clear of. Knowing that the Titanic had struck ice, of course, I had to take extra care to keep clear of anything that might look like ice. I doubled my lookouts and took extra precautions. We were all on the qui vive. We were passing icebergs on every side and making them ahead and having to alter our course several times to clear the bergs.*

Not once did he slow down, but kept his ship thundering through the dead calm seas. By this time there had been no radio signal received from the *Titanic* for 45 minutes. Rostron was not sure if this meant the liner had lost sufficient power to send radio signals, but he worried she might have sunk, which she had.

At about 3 am, Captain Moore of the *Mount Temple* had a minor fright. Coming out of the darkness towards him was a schooner, probably a fishing boat, working the rich cod stocks of the Grand Banks. The schooner sounded its foghorn to alert the freighter thundering towards it. Moore heard the horn, spotted the sailing ship and slammed his ship hard to port, passing the schooner and clearly seeing its green starboard light. The schooner had been coming from just north of west on a course that meant it had travelled from slightly to the north of the *Titanic*'s reported position.

Soon after 3 am, Rostron on the *Carpathia* gave orders that the ship's night signals should begin to be launched. Cunard's night recognition lights consisted of a blue light on the masthead plus two roman candles throwing blue stars to a height of 150 ft (46 m) or so. He estimated that if the *Titanic* was still afloat men on the top deck would be able to see his blue stars as they peeked up over the horizon. Later Moore decided to launch his own distress rockets to make his ship even more visible. He was hoping to see an answering night signal from the *Titanic*, but none came. The suspicion that the ship had sunk was becoming more of a certainty.

Then the men on the *Carpathia* saw the lights of a steamer ahead of them, but the vessel was too small to be the *Titanic*. Rostron recalled: 'We saw masthead lights quite distinctly of a steamer, and one of the officers swore he also saw one of the sidelights, the port sidelight. It was about 2 points on my starboard bow, that would be a bearing of about 30° North.' The stranger was coming down from the north about 15 to 20 miles (24 to 32 km) east of the scene of the sinking.

Meanwhile, Captain Moore of the *Mount Temple* was running into difficulties. 'At 3.25 am we began to meet the ice. We were passing it on our course. I immediately telegraphed to the engine room to stand by the engines. We doubled the lookout. I had one man in the crow's nest and we put an extra man on the bridge and I put the fourth officer forward in the bows to report if he saw any ice coming along that was likely to injure us.' Having taken steps to protect his own ship from striking an iceberg, Moore restarted the engines, but he now proceeded at a reduced speed so that he would be able to take evasive action if an iceberg did loom up out of the darkness.

Moore estimated that he was now about 15 miles (24 km) from the position given by the *Titanic*. His radio operator had heard nothing from the *Titanic* for some time, which might have indicated that the liner had sunk – as indeed it had. However, at a distance of 15 miles the *Mount Temple* would have been within sight of signal rockets bursting high in the sky. Neither Moore nor his doubled lookouts saw anything.

By 3.50 am the *Carpathia* was getting close to the *Titanic's* reported position. Captain Rostron slowed down so that his doubled lookouts had more of a chance of seeing any wreckage floating in the dark water. Rostron thought that he saw a flash of green light far ahead, but it was down on the water, not high up as it should have been if the *Titanic* was responding to his flares. It must, Rostron thought, have come from a lifeboat. Rostron remembered:

At four o'clock I considered I was practically up to the position and I stopped, at about five minutes after four. We could not see anything. In the meantime I had been firing rockets and the company signals. At five minutes past four I saw a green light, and I was going to approach it on the port bow, but just as it showed I saw an iceberg right ahead of me. It was very close, so I had to put my helm hard-a-starboard and put her head around quick.

The green light turned out to be a flare lit by Fourth Officer Boxhall in Boat No. 2. He had seen the liner driving north at full speed and had worried that the boats might be run down. Boxhall steered the boat so that it came up alongside the *Carpathia*, then supervised the survivors going up ladders or being lifted up in chairs and slings. Finally he scrambled up himself. He had one last duty to perform and headed for the *Carpathia*'s bridge. There he informed Rostron that the *Titanic* had gone down. There were, Boxhall said, 20 lifeboats that needed to be found. Rostron turned to the grim task of finding the boats in the ice-strewn sea.

On board the Russian ship *Birma*, Captain Stulpin was meanwhile getting things organized. 'At 4 o'clock the crew was called out and we began to prepare all things necessary for the reception of the shipwrecked people.' Among the preparations begun was the cooking of huge saucepans of hot soup.

At 4 am Stone was replaced as officer of the watch on the stationary *Californian* by Chief Officer Stewart. As usual Stone chatted to Stewart about weather conditions, and then told him about the steamer that had approached, then gone away again and about the firework rockets seen over the horizon. Stewart was uneasy about what he was told. Ships did not fire rockets at sea for no reason. The white rockets and the time over which they had been fired did not sound to Stewart like company recognition signals. After half an hour on watch, he went below to wake up the radio operator Evans and asked him to switch on the radio set to see if there was any news.

The *Californian* joins the mission

Almost at once, Evans picked up a signal from the German ship *Frankfurt* reporting that the *Titanic* had sent out a distress signal and the position from which it had been sent. The position was not the same as that from which the rocket signals had been seen – it was further west – but clearly immediate action was called for. Stewart roused Captain Lord, and the *Californian* was on her way. Lord decided to nose slowly through the ice field as the reported position of the *Titanic* was on the far side, then to steam south at full speed.

By about 4.20 am it was light enough for Rostron on the bridge of the *Carpathia* to see several miles around his ship. The entire sea surface was littered with icebergs, growlers and field ice. Here and there were lifeboats splashing towards him. It also allowed the survivors in the four boats lashed together by Fifth Officer Lowe before he set off to rescue people in the water to see a bizarre sight. About 800 yd (730 m) away, some 30 men were standing up on what looked like a ship's funnel. It was, in fact, the overturned collapsible Boat B on which stood Second Officer Lightoller, Joughin the baker, radio man Bride and others.

Lowe had taken most of the good rowers with him, but Boat No. 4 still had Quartermaster Perkis and No. 12 had Able Seaman Fred Clench. When they heard a shrill whistle blast coming over the water, they realized that an officer was among the standing men and that he was calling them. Perkis and Clench asked for volunteers to row, then cast off from the other boats and headed for Lightoller, still blowing desperately on his whistle.

The two boats came alongside the overturned boat only just in time. The growing waves were now washing over the keel of the collapsible, sloshing around the men's ankles. One by one the survivors scrambled into the lifeboats. Lightoller was last to leave, jumping into Boat No. 12. He then reorganized the men and women in the two boats, producing two good rowing crews between them and set off towards the *Carpathia*, visible some 4 miles (6.4 km) away to the south. The exhausted survivors from Boat B were given dry coats and other oddments donated by the dry folk in the two lifeboats. Lightoller was given a rather incongruous lady's hood and cape handed over by a teenage girl, Madeleine Mellenger.

The *Mount Temple* pushes on

The *Mount Temple* was meanwhile continuing to push east towards the *Titanic*'s given position. At 4.30 am, Moore reckoned he had reached the spot and stopped his engines. Moore had his crew on deck by this time and instructed them to look about for lifeboats, wreckage or flares set off by survivors in boats. They could see nothing, nor could they hear anything. An eerie silence and stillness drifted over the ocean.

The *Mount Temple* was not quite alone out there. Off to the north Moore and his lookouts could see the lights of a small merchant steamer. She was heading slowly south on a course that would take her across the *Mount Temple*'s bows 2 miles (3.2 km) or so off. Moore pushed his ship slowly to the east, but then came across what he described as

a large ice pack right to the east of me, right in my course. It extended as far as the eye could see north and south. It was about 5 miles [8 km] wide east to west. There were bergs interspersed in the pack and growlers. I should say there were between 40 and 50 icebergs within sight. The largest one was, I should say, about 200 feet [60 m] high. We searched around to see if there was a clear place we could go through. Of course, I reckoned I was somewhere near if not at the Titanic's position that he gave me. I went south and searched for a passage to get through this pack because I realized that the Titanic could not have been through that pack of ice. I could find no way through. I slowed down and stopped there. I had a man hauled to the very masthead on a rope and he could not see any line through the ice at all.

By this time it was after dawn. Moore took a sighting on the sun and calculated that he was now 4 miles (6.5 km) south and slightly east of the position given by the *Titanic*. Moore concluded that the position given him by the *Titanic* must have been incorrect. The big liner could not have made her way through the pack ice and so must have sunk somewhere to the east of it.

The ship seen coming from the north could now be seen quite plainly from the *Mount Temple* as it steamed away to the south, apparently also looking for a way around the ice. Moore later described her as being a tramp steamer, that is a ship not following any designated schedule or route but which picks up whatever freight is available and takes it wherever is required. Moore said of her:

She was a ship of 4,000 or 5,000 tons. I did not get her name, but I think she was a foreign ship. She was not English because she did not show her ensign. We were trying to signal with her, but had no communication

with her. Her funnel was black with a device of some
kind in a white band near the top. She had four masts.

The *Frankfurt* is denied by the ice

Not long after dawn another ship hurrying to the rescue
was brought up short by the ice barrier. This was the
German liner SS *Frankfurt.* Her radio operator had not
tried to contact the *Titanic* after his last question had
been slapped down, but he had been monitoring the
signals. Faced by the great mass of field ice and icebergs,
the *Frankfurt,* like the *Mount Temple* before her, turned
south and steamed slowly, looking for an opening.

At 7.30 am, the *Frankfurt* was followed by the Russian
Birma, now thundering along at her top speed. Captain
Stulpin recalled:

About 7.30 am we arrived at the above mentioned
scene of the wreck. There we saw some immense
icebergs to the east and as far as the eye could reach
there lay pack ice so that it was out of question to

*proceed through. It was quite clear the ship could
not be at that spot. We received information by
telegraph [he meant radio] that the Carpathia was
picking up boats.*

In the meantime the *Carpathia* had been picking up
survivors from lifeboats. As each person clambered
up or was lifted on board, they were led to the saloon
where they were given food and a hot drink. Mostly
they went in silence, though many women were crying.
A few desperately asked if relatives or friends were on
board; the stewards did not know and led the survivors
to the saloon. There was one exception, a middle-aged
man in an overcoat and pyjamas. He refused to be led
below but stood with his back to a bulkhead watching
the sad procession of survivors troop past him. When
a steward tried to pull him forcibly towards the saloon,
the man shook him off and glared. 'Leave me alone,' he
muttered. 'I am Ismay, I am Ismay.' Then he went back
to watching the survivors arrive.

When Boat No. 4 came alongside it was accompanied by a big, black Newfoundland dog that had been swimming alongside it ever since the *Titanic* had gone down. The dog barked continuously, but was too big and heavy for anyone to lift it up out of the water. Then Jonas Briggs, one of the *Carpathia*'s crew, sent down a canvas sling that he managed to get under the dog's belly and so lift it aboard. The dog was later identified as having been the pet of First Officer Murdoch. Briggs adopted the dog and kept it thereafter.

The final survivors board the *Carpathia*

The last of the lifeboats to reach the *Carpathia* was Second Officer Lightoller's new vessel, Boat No. 12. It pulled alongside at 8.30 am. Like Boxhall before him, Lightoller was the last to leave his boat and scramble up to the decks of the *Carpathia*. And like Boxhall before him, he headed straight up to the bridge to report to Captain Rostron. Rostron was worried. Boxhall had told him there were 20 lifeboats to find, but he had accounted for only 19. Lightoller explained about the overturned collapsible. That made 20. Rostron asked if it were likely that anyone was still alive in the water; Lightoller thought it was not given the freezing temperatures.

Nevertheless, Rostron wanted to make certain. He nosed his ship slowly towards the main field of wreckage just in case someone was clinging to some wooden flotsam. For an hour or more the big ship prodded about while men shouted and blew whistles to alert anyone with life still in them. There was no response.

Meanwhile, the chief doctor on the *Carpathia* had been sent by a concerned steward to look at the man calling himself Ismay who refused to go below. The doctor recognized the man to be the Mr Ismay who was the head of the White Star Line. Like the steward before him he tried to get Ismay to go to the saloon.

'No. I really don't want anything,' Ismay insisted. 'If you leave me alone I will be much happier.' In the end the doctor persuaded Ismay to go to a nearby cabin. Ismay sat on the bunk staring blankly at the wall and said nothing.

The rescue mission is called off

Cottam, *Carpathia*'s radio operator, had meanwhile been sending out radio signals far and wide to tell other ships still steaming at full speed towards the *Titanic* that the emergency was over. Everyone who could be saved had been saved. The *Virginian*, *Baltic*, *Caronia* and others turned about and resumed their courses.

The *Californian* heard all this while pounding south along the western side of the ice field at her top speed of 12 knots. She had cleared the ice field at about 6 am and turned south. At around 7 am she reached about the position given for the *Titanic* by the *Frankfurt*. Only one ship was in sight, the *Mount Temple* some distance to the south. The *Californian* steamed on. She was now in radio contact with the *Carpathia* and knew that she was just east of the ice field, picking up survivors.

After being passed by the *Californian*, Captain Moore of the *Mount Temple* also headed south again to look for a way through the ice. A ship was now sighted some miles to the east on the far side of the ice field. Moore did not recognize her, but it was the *Carpathia*. He had not gone far when his radio operator came up to the bridge with a message from the *Carpathia*. The radio man on the *Mount Temple* had been in touch with the *Carpathia*, which had already reported that it was on the site of the sinking and was picking up survivors. This new signal told Moore that 20 boatloads of survivors had been picked up and that there were no more survivors. It ended: 'No need to stand by. Nothing more can be done.'

Captain Moore told his crew to stand down. The lifeboats were swung back in and recovered, the stewards went about their normal duties rousing the passengers for breakfast and a fresh course was plotted for Montreal. The *Mount Temple* swung back west and steamed off.

The *Californian*, however, pushed on. When her lookouts spotted the Cunarder on the far side of the ice field, Captain Lord turned towards her and for the second time that morning gingerly edged through the growlers and field ice. Lord estimated that he had travelled about 25 miles (40 km) that morning, and that he found *Carpathia* more than 20 miles (32 km) south of where he had started from. Lord remembered:

> *I think the Carpathia was taking the last boat up when I got there at 8.30 am. I saw several empty boats, some floating planks, a few deckchairs and cushions, but considering the size of the disaster, there was very little wreckage. It seemed more like an old fishing boat had sunk. There were no bodies.*
>
> *We went in circles over a radius, a big circle. Then came around and around and got back to the boats again where I had left them. I was practically surrounded by icebergs, the ones to the south-east were much larger than the ones to the west. I suppose the largest was about 100 to 150 feet [30 to 45 m] tall.*

Captain Rostron was keen to escape the scene with the survivors he had on board, and planned to take them to New York.

At about 11.30 am, Captain Lord of the *Californian* was still looking for any last survivors, but had just about given up hope. He saw a ship nosing warily towards him through the sea ice. It was the German liner *Frankfurt*. As before there were some language problems, but Lord finally managed to get the *Frankfurt* to understand that there was nothing more to be done. The German ship bore away and resumed her original route.

Once underway towards New York, *Carpathia* had to pick a way through the ice field. She was no sooner on the far side of the ice than she came across the Russian ship *Birma* still trying to get to the scene of the disaster and offer help with her saucepans of hot soup. 'About 12.15,' Captain Stulpin of the *Birma* recorded, 'we met the SS *Carpathia* going at full speed to the west and exchanged flag signals with her. I asked if everybody was saved and whether we could help in any way. We were told to stand by. I ordered again the enquiry to be made is there any use in our remaining to search further.' Being told there was not, Stulpin went on his way. Presumably his own crew had the soup.

As the *Carpathia* steamed west, Rostron held meetings with Lightoller to discuss the disaster. He would have liked to have talked to Ismay, but the White Star chairman would not come out of his room and would not talk to anyone. The *Carpathia*'s doctor judged that he was 'in a terrible nervous condition'. The doctor sent a survivor named John Thayer, who was acquainted with Ismay, to try to talk to him. Thayer recalled that:

> *He was seated in his pyjamas on his bunk staring straight ahead and shaking all over. My entrance apparently did not dawn on him. Even when I spoke to him and tried to engage him in conversation he paid absolutely no attention and continued to look ahead with his fixed stare. I have never seen a man so completely wrecked. Nothing I could do or say brought any response.*

The death toll becomes apparent

Together, Lightoller and Rostron's purser drew up lists of survivors. Of the 2,223 people on board the *Titanic* when she hit the iceberg, only 706 had been saved. As they worked over the figures it became apparent that there had been clear differences between the different categories of people. Of the first-class passengers 60 per cent had been saved, of second class just 42 per cent and of third class only 24 per cent. The crew had fared even worse, with only 23 per cent surviving the disaster. Of the men on board only 20 per cent had got away, but 75 per cent of women and children had survived. The biggest loss rate was among the orchestra and the team from Harland and Wolff, all of whom had been lost. The restaurant staff did little better with a death rate of 96 per cent. Of the ship's officers, Captain Smith, Chief Officer Wilde, First Officer Murdoch and Sixth Officer Moody had perished.

Meanwhile, the survivors were sorting themselves out. Sir Cosmo Duff Gordon, as he had promised, wrote cheques for £5 each for the crewmen who had been in his lifeboat. Charlotte Appleton found that her uncle, Charles Marshall, was on board the *Carpathia*. Mrs Ruth Dodge and her son Washington had got away in Boat No. 5 and was convinced her husband had died until she saw him on the *Carpathia* – he had escaped in Boat No. 13. Most other women were less lucky, finding that missing husbands, sons and brothers were nowhere to be found.

There were two young children who were a total mystery. These were the two boys handed to Lightoller as he was getting collapsible Boat A away at the last minute. Neither of the toddlers spoke any English at all. One passenger thought they had been travelling with a man called Hoffman, but could not be certain.

In the hours and days that followed, Cottam of the *Carpathia* and Bride of the *Titanic* sent out a steady stream of radio signals to the shore stations. Most of these were lists of survivors and personal messages from those survivors to their families. Others were business messages signed by Joseph Bruce Ismay, but mostly written by Lightoller. Rostron ordered that absolute priority be given to these messages and that no others should be answered or even acknowledged. As he suspected, *Carpathia* was inundated by messages from the press wanting details and stories for their newspapers. They were all ignored. The newspapers were irritated by this and lambasted both the crew of the *Carpathia* and the Marconi company which operated the radio. Rostron was always convinced he had been behaving correctly.

One of Ismay/Lightoller's principal concerns was to get the crew of the *Titanic* home to England as quickly as possible. Several messages were sent to the White Star offices in New York asking them to arrange passage home at the first opportunity, even suggesting delaying the departure of ships to allow the crew to transfer directly from the *Carpathia*. It was not to be – Senator William Smith of Michigan had persuaded the US Senate to hold a formal inquiry into the loss of the *Titanic*. The officers, crew and passengers were going to be wanted in Washington to give evidence.

CHAPTER 10

THE INQUIRIES

Senator William Smith, who called for and led the Senate inquiry into the *Titanic* sinking, was a man with an axe to grind. He had been born into poverty and had built his political career on supporting small businesses and farmers against big companies and trading combines. In particular, Senator Smith had tangled several times with the banker and financier J.P. Morgan. And if most people in Britain had not yet realized that the White Star Line was owned by J.P. Morgan, then Senator Smith had. He knew that Morgan had tried, and almost succeeded, in cornering a near monopoly in the Atlantic passenger trade. He sensed in the sinking an opportunity to expose to public view Morgan's shipping interests and what he suspected to be short cuts on safety that had resulted. In any case, he was up for re-election that year and the publicity would not hurt him one bit.

Although Senator Smith was intelligent, quick-witted and believed himself to be working in a virtuous cause, he knew nothing about the sea or about ships. Time and again witnesses at his inquiry found themselves having to correct him when he got muddled. He thought the bows of the ship referred to by one witness were different to the head mentioned by another when the words meant the same thing. He asked another witness what an iceberg was made of. 'Ice', came the reply.

There can be no doubt that the crew of *Titanic* deeply resented being forced to stay in the USA for Senator Smith's inquiry. Not only did they consider many of his questions to be foolish, but they were keen to get home to their families. Added to that was the knowledge that because the *Titanic* had been a British-registered ship the official inquiry into the sinking would take place in Britain. Having given exhaustive evidence in the USA, the officers and crew knew they would have to do it all over again in Britain. And if that was not enough, Smith encouraged passengers who knew next to nothing about how ships operated to speculate at length on the causes of the disaster. It all made for great newspaper copy for the press, but did not get the world much closer to understanding what had occurred. The tetchiness of some witnesses is more than clear in the transcripts of the inquiry. Lightoller rarely said much at all. His answers to questions consisted largely of 'Yes', 'No' or 'I could not say'.

However, Senator Smith's inquiry did serve to uncover a number of features about the sinking that the more formal and technical hearings in Britain might have missed.

The feature that most excited the American press was the behaviour of Joseph Bruce Ismay, chairman of the White Star Line, and how it contrasted to that of the famous Americans on board. The elderly Straus couple had refused places in a boat and were last seen devotedly cuddling each other. John Jacob Astor had put his wife into a boat, lifted in a lone third-class child and then stepped back to make room for others. Benjamin Guggenheim had dressed up in formal evening wear to meet his death like a gentleman. But Ismay had grabbed a place in a boat and got away safely while women and children were left on board.

When Ismay appeared before the Senate inquiry he had recovered some of his composure, but not much. He was hesitant, vague and prone to long pauses. Sometimes he would start an answer, stop to correct himself and then get the facts wrong anyway. When it came to the question of why he had got into a lifeboat, Ismay stuck to the story that Chief Officer Wilde had told him to do so. He presented himself as just another passenger, but one who was luckier than most. The American press, however, did not think he was just another passenger. As the chairman of the shipping line that owned the vessel, they cast him in a role similar to that of the captain. They argued that his duty was to ensure the safety of all others first, and only then to look to himself. According to the customs of the sea they were wrong and Ismay was right, but the public and the inquiry saw things as the press did. Ismay's refusal to accept he had done anything wrong went down very badly.

The other question that Senator Smith concentrated on was whether or not Ismay had forced Captain

Smith to drive the *Titanic* faster than it was safe to do. Undeniably, Ismay had told many passengers on the Sunday that he had discussed the question of speed with Captain Smith and that it had been decided to fire up all the boilers and drive the ship at full speed for several hours on the Monday. Ismay had seemed proud of this fact and was urging the passengers to get ready to see something special. Moreover, it was proved that Ismay had been shown at least one of the ice warnings received that afternoon. And he was known to have spoken to Captain Smith for some time on the evening of the disaster.

Ismay quickly refuted suggestions that the ship had been trying to win the Blue Riband. He pointed out that the White Star Line sold itself on luxury and left Cunard to corner the market in speed. Moreover, *Titanic* was simply not fast enough to win the coveted prize. And every officer called as a witness stated clearly that it was the captain who had command of a ship while it was at sea. It was a custom of the sea that was rigidly adhered to by all. The captain was responsible for everything and so he made the decisions. Captain Smith, they testified, was an old hand close to retirement and a strong character. He would never have agreed to anything he thought unsafe.

Nevertheless, there remained the fact that many passengers had the impression that it had been Ismay's idea to run at full speed on the Monday and that Smith had agreed to the chairman's suggestion. What else might have been suggested that fatal evening?

Ismay's story did not change when he appeared at the British inquiry, though he delivered with rather more assurance. Having finished giving evidence, Ismay slipped away and almost vanished. He resigned from the White Star Line and most of his other positions, bought a remote sporting estate in Ireland and rarely left it. He died in 1937.

Did the 'mystery ship' exist?

Another curiosity dug up by the American inquiry

related to the 'mystery ship' that had been seen by several of those on the *Titanic* as she was sinking. An American newspaper had carried a story quoting a member of the crew on board the *Californian* saying that he had seen the *Titanic* sinking, but that the commander of his ship, Captain Lord, had done nothing to help the stricken liner. It was a sensational story, though it was undermined by the fact that the crew member in question, Ernest Gill, had been paid $500 for his story (an amount which came to more than a year's wages), and by the fact that Gill quickly left the scene after getting his money.

Nonetheless the accusation was serious. If true, Lord had broken one of the most important customs of the sea, and with it a basic humanitarian demand. He had failed to go to help those in peril.

By the time Captain Lord was questioned by Senator Smith, he knew that he was potentially in a lot of trouble. His evidence was obviously guarded as he carefully thought about his answers. Generally, Lord answered questions about facts – such as times, positions and bearings – clearly and succinctly without adding any extraneous detail. When asked to speculate on the actions or motives of others, Lord refused to be drawn. 'I could not say', 'I have no experience of that' and 'I do not know' were heard often.

When it came to whether the ship seen from the *Titanic* was the *Californian*, and whether the ship seen from the *Californian* was the *Titanic*, Captain Lord was emphatic. The ship he had seen was a small freighter, not a big liner. It had steamed off without showing any signs of distress. The fireworks seen were more like company recognition signals than distress signals. In any case, the log of the *Californian* put her position as some 20 miles (32 km) north of the position broadcast from the *Titanic*. From that distance the ships would have been over the horizon from each other and so out of sight. Although he did not state it, Lord's contention was that there had been a third ship that had been lying between the *Californian* and the *Titanic* for most of the night, but had moved off before dawn.

At the end of his inquiry, Senator Smith reached a number of conclusions that he announced on 28 May. He found that the *Titanic* had been going too fast in an area where ice was known to be present, that the evacuation of the ship left a lot to be desired and had been disorganized at times, and that those in the lifeboats had shown a shocking indifference to the plight of those in the water by not going back to rescue them. He recommended that the use of radio at sea be reformed to ensure that radio sets were manned 24 hours a day. He found that Ismay had not ordered Captain Smith to go too fast, but that his presence on board might have encouraged the captain to try for a speedy passage.

Finally, Senator Smith came down hard on Captain Lord.

'Captain Lord deluded himself about the presence of another ship between himself and the Titanic. There was no such ship. He bore a heavy responsibility.'

In the meantime the press and public had been following their own course. There was much criticism of the way in which news of the disaster had arrived. Most of the news had come by way of Cape Race radio base which passed on the messages flying back and forth between the *Titanic* and the various ships rushing to the rescue. Given the distance the signals had to travel, the vagaries of atmospherics and the variable quality of the ships' equipment, there were inevitable misunderstandings. The operators at Cape Race passed on all messages over the telegraph lines without taking the time to confirm the news. Thus it was reported that the *Titanic* was under tow, when in fact a quite different freighter had been taken in tow hundreds of miles away. The *Virginian* was reported to be alongside the *Titanic* taking off passengers, when in fact she was offering to take survivors off the *Carpathia*. Demands for the Marconi company to improve its procedures were loud. Marconi promised to revise the operating instructions and to train their men more rigorously.

Other messages were more straightforward. As news of the sinking was received, Kaiser Wilhelm II of Germany sent a telegram to Britain's King George V expressing his deepest sympathy to the British people and offering the assistance of the German government. Messages from other countries followed, including from the Russian Tsar, the Mayor of Rome, the Hungarian parliament and the town council of Sydney, Australia.

Fundraising for the bereaved

Also rising in volume were demands for efforts to be made to help the widows and orphans left by the disaster. On 18 April, the Lord Mayor of London opened a fund at the Mansion House and began organizing fundraising events. Newspapers throughout Britain carried messages asking any dependents of crew members who had been lost to write to the Lord Mayor before 1 July giving details of their names, addresses, the man lost and what his pay had been. The fund was eventually divided up accordingly.

On 19 April the *Daily Sketch* newspaper started what they called the Save Our Shillings fund, inviting every reader to send in one shilling and kicked it off with a donation of 1,000 shillings by the newspaper itself. 'If you can send no more, send a shilling,' the editorial read. 'If you can afford more, all the better.' In all the fund raised the considerable sum of 18,361 shillings – more than £918.

Individual groups did their bit for those to whom they had a link. An organization of amateur radio operators in Argentina gathered a few pounds that they handed to the British ambassador for the family of Jack Phillips, the radio operator who had lost his life. Seven British orchestras combined their finest talents to lay on a special concert on 24 May, the proceeds of which went to the families of the *Titanic*'s band. Hundreds of other efforts raised large sums to help the bereaved. The townsfolk of Barmouth in North Wales clubbed together to buy a watch for Fourth Officer Harold Lowe, who had been born there, to replace the one he lost in the wreck.

The disaster, meanwhile, had a strong but temporary effect on the stock markets on both sides of the Atlantic. Shares in shipping companies slumped in price as the news of the sinking arrived, but picked up again after a few weeks.

One of the first mysteries to be sorted out was that of the two toddlers travelling with a man named Hoffman. They were found to be the missing children of a Mrs Navatril. She had fallen out with her husband, who had then left with the children, leaving no word of where he had taken them. Mrs Navatril informed the police who began a search and finally found that Mr Navatril had booked passage on the *Titanic* as Mr Hoffman. The children were returned to their mother.

Other odd stories surfaced. A Mr J.C. Middleton, an American railway manager, had a ticket for the *Titanic* but cancelled it three days before the ship left port. 'I dreamed that I saw a ship capsized in the ocean,' he said later. 'I saw a lot of passengers struggling in the water. I decided to cancel my passage.' Just as odd was the experience of Mrs Gracie, the wife of Colonel Archibald Gracie who survived on the upturned collapsible boat. She was asleep in her sister's house in New York on the night the *Titanic* went down. Suddenly something woke her up. 'What is the matter?' she called out into the darkness. A voice came from nowhere: 'On your knees and pray', it said. She did so, grabbing a prayer book that fell open at the hymn 'For Those in Peril on the Sea'. She stayed awake the rest of the night worrying about her husband.

The British inquiry

The focus of interest had meanwhile moved to Britain for the British inquiry. It opened on 2 May, having been delayed by the fact that so many of the witnesses were stuck in Washington appearing at the American inquiry. The British inquiry was organized by Lord Mersey. Mersey had been born simple John Bigham, son of a moderately successful Liverpool merchant. He opted for a career in law, but remained in Liverpool and specialized in shipping cases. He became a judge in 1897, but retired in 1909 after a heart attack. By 1912 he had recovered from the heart attack and although he did not want to return to full-time work, he was looking for a job of some kind. His knowledge of shipping and the high regard in which he was held made him the obvious choice for the inquiry.

Unlike the questioning led by Senator Smith, that under Lord Mersey was much more concerned with technical matters. While Senator Smith's inquiry was carried out in the full glare of media attention and at times seemed designed for sensationalism to get Smith on the front pages, the British inquiry was a much more subdued and dull event. There were endless questions about ship-board procedure, the testing of the ship and its equipment, the layout of the ship, the actions of the officers and men. For day after day it dragged on with only the odd spark of excitement to get the attention of the press.

One such event was the evidence relating to Sir Cosmo

Duff Gordon. This wealthy baronet had been the man who gave all the seamen in his lifeboat a cheque for £5 to help them buy new kit. The way this was reported in the newspapers made it seem as if Gordon had paid the men not to go back to rescue survivors. It was made to appear as if the crew in Boat No. 1 had wanted to row back to the wreckage after the ship had gone down, but that Gordon had not wanted them to do so and solved the argument by paying them.

The story caused a sensation, and the Gordon family was outraged. Sir Cosmo wrote to Lord Mersey asking if he could appear to give evidence under oath and so dispel the story. The behaviour of a single baronet was not really the business of the inquiry, but Sir Cosmo was a rich and influential man so Lord Mersey agreed. The cream of London society turned out for the day when Sir Cosmo appeared and the press gallery was packed. It all turned out to be an anticlimax. The evidence of Sir Cosmo and others in the boat was unremarkable. They had failed to go back to help, but so had others. He had paid the men, but not until much later. The story fizzled out.

Lord Mersey heard in all some 94 witnesses discussing, in the course of 27,000 questions, the design of the *Titanic*, how captains should behave in the presence of ice, what the precise duties were of the various officers and crew members, what was standard emergency procedure on White Star Line ships, what types of davits were used to launch the boats, the composition of the Socket distress flares, the wording of individual radio messages. The hearings dragged on, day after day, until early July. By then the interest of the general press had moved on elsewhere and the evidence was barely reported. Reading the transcripts today, it is hardly surprising.

The final report, published on 30 July, was covered extensively and, by the shipping trade press, in great detail. Lord Mersey's final report ran to thousands of words, and he made dozens of recommendations, some of which were highly technical and very detailed.

Some individuals, such as Second Officer Lightoller and Captain Rostron of the *Carpathia*, came in for praise while others, Captain Lord of the *Californian* among them, found themselves criticized. On the whole, however, Lord Mersey did not make it his business to apportion blame or responsibility. It was his job to make recommendations that would prevent such a terrible disaster from happening again.

Lord Mersey noted that the practice of mail steamer captains was to press on at speed in almost all circumstances and that Captain Smith had merely been following this established habit. He pointed out, however, that the RMS *Titanic* had not been an old-style mail steamer and should not have been treated as one. In future captains should not steam at speed when ice was known to be about. Captain Smith could not be blamed for what he had done, but any captain who repeated his actions could and would be blamed.

Similarly with the question of lifeboats. Harland and Wolff and White Star had followed the rules from the

Board of Trade when it came to how many lifeboats the *Titanic* should have on board. The Board of Trade had published rules based on then current practice and the best advice it could find. Nobody had done anything illegal or improper, but events had shown that they had been gravely mistaken. New rules for lifeboats were suggested.

In fact, public opinion was already having an impact. Passengers were refusing to book on ships that did not have enough lifeboats for everyone on board, so the shipping lines were already loading more boats on to their ships before Lord Mersey reported. Even more direct was the fact that the crew of the RMS *Olympic*, the *Titanic*'s sister ship, had refused to put to sea until enough boats for everyone on board had been loaded. The ship eventually sailed 20 days late.

Lord Mersey recommended that lifeboat drills be carried out soon after a ship left port, that all liners no matter what their size should have radio sets manned round the clock, and laid down technical specifications for what types of radio they should have. He laid out detailed notes on the construction of future large liners, covering such subjects as watertight compartments, signage for the use of passengers and layout of decks. He made recommendations about the types of insurance shipping companies should be forced to take out for passenger ships. Lord Mersey suggested that the world's maritime powers should get together to agree standard rules and procedures for the ways in which ships and shore stations should react to emergencies. Two years later that came to pass with the International Convention for the Safety of Life at Sea. Mersey covered just about every aspect that could possibly have a bearing on the future safety of passengers at sea. The report was well received and considered to be the last word on the subject. It would, everyone thought, guide the future conduct of the passenger trade for decades to come.

But this was not to be the case. Just two years later the world was plunged into the First World War. Britain was at war with the same Kaiser Wilhelm who had offered his sympathy to the British people and with the Hungarian parliament that had sent kind words of support. The horrors of the trenches and the vast casualties marked a watershed in society as much as in history.

The rigid class barriers that had seen the *Titanic* so firmly divided into separate areas collapsed. The differences in survival rates between first class and third class had passed almost unnoticed in 1912, but soon came to appear deeply unfair and anachronistic. Some suggested that they reflected what was seen as the bad old days. There were even efforts made to try to demonstrate that first-class passengers had been given preferential treatment or that those in third class had been locked down below to drown.

Others looked back to the *Titanic* as the embodiment of better times when heroic self-sacrifice and devotion to duty was the norm. The behaviour of the ship's band was held up as being exemplary, as it was. The way in which the crew stuck to their duty even though almost certain death awaited them was hailed as showing all that was best among the old society before the Great War.

The sinking of the *Titanic* did much to destroy the faith of Western society in technology. For more than a century the wonders of the industrial revolution had seemed to produce nothing but benefit for humanity. More food, better housing, piped and purified water, gas lights and gas heating, better and cheaper clothing, more job opportunities and vastly improved transport links. Each new advance had been hailed by international fairs and enthusiastic newspaper reports, with promises of even better to come.

But now the latest wonder, the great superliners that would ply the world's oceans and tie humanity together in a common brotherhood as never before, had been brought crashing down in disaster. The supposedly unsinkable ship had sunk. The Bishop of Winchester preached a sermon, the main message of which was that the iceberg had every right to be there, but that the *Titanic* had not. People began to lose their unquestioning faith in technological advance. The promises made by scientists and industrialists were treated with more scepticism and subjected to more study before being accepted.

Coming so soon before the advent of the First World War and its horrors, the event seemed to symbolize the death of a bygone era. The world of Edwardian splendour and self-confidence had gone down with the *Titanic*. But though the ship was gone it was not forgotten.

CHAPTER 11

BACK TO THE *TITANIC*

The sinking of the *Titanic* continued to exercise a fascination over the years. It was the worst peacetime maritime disaster in history and had a vast number of romantic features: the luxury of the ship, the romance of a maiden voyage, the beautiful starlit night, the heroism of those who died, the tragedy of the bereaved, the slow-moving drama as the waters moved through the gigantic ship, dragging her to her doom.

And then there were the unanswered mysteries of that terrible night. Witnesses argued with each other as to whether the ship had split in two just before she sank. There were differences of opinion over what orders had been given in the evacuation, but the greatest of these was the identity of the 'mystery ship' that had not come to the aid of the sinking liner. Both the American and British inquiries had implied that the ship had been the *Californian* commanded by Captain Lord, but had stopped short of actually saying so. Neither Captain Lord nor the Mercantile Marine Service Association, the trade union for merchant ship offices, accepted that verdict.

Lord and his supporters pointed out a number of reasons why the mystery ship could not have been the *Californian*. First, the ship seen from the *Titanic* had moved towards the liner, getting as close as about 4 miles (6.5 km). It had stayed there for a while and then moved away. The *Californian*, however, had been stationary all night.

Second, the ship seen from the *Californian* had come into sight at around 11 pm, stopped at about 11.50 pm and moved out of sight about 2 am. If the ship seen from the *Titanic* was the *Californian* it would be reasonable for the men on the *Titanic* to have seen the *Californian* at about the same time, but they did not. The mystery ship was first seen about 12.30 am and was last seen some time after 2.20 am.

Third, the Socket distress signals fired by the *Titanic* were designed to be visible from a distance of up to 20 miles (32 km), depending on visibility, and to explode with a blast audible up to 6 or 7 miles (10 or 11 km) away. Nobody on the *Californian* heard a thing.

Fourth, the Socket signals were fired from the *Titanic* between 12.25 am and about 1.40 am. However, the rockets seen from the *Californian* were visible from about 1.30 am to 2.10 am.

Rather more complex was the issue of where, precisely, the *Californian* was while the *Titanic* was going down. Captain Lord maintained that his ship was at 42° 5' North 50° 7' West, some 20 miles (32 km) north of the *Titanic*'s reported sinking position of 41° 46' North 50° 14' West. From that position the liner would have been well over the horizon and the Socket signals at the very limit of their visibility. However, Lord's position was not beyond dispute. He had not taken a firm sighting at dusk but had taken his last sighting to fix a firm position at dawn. Since then he had been estimating his position by dead reckoning based on his speed and direction of travel. Lord was considered by his colleagues to be above average at dead reckoning, but even so he would have been doing well to be within 4 to 5 miles (6 to 8 km) of his true position after a full day's travel.

Finally, Lord's supporters pointed out that there were other ships in the area that night. The *Carpathia* had passed a steamer coming towards them as they ploughed north to the rescue. The *Mount Temple* had seen a tramp steamer at dawn heading south on the western side of the ice field. Great efforts were made at the time to identify these ships, but with no result. There is no proof that either of them was the mystery ship, but there is no evidence that they weren't. At any rate, the *Californian* was not the sole suspect.

The *Titanic* is found

For years the arguments raged back and forth. Then in 1985, long after Captain Lord had died, came spectacular new evidence from a quite unexpected quarter. The wreck of the *Titanic* was found on the floor of the Atlantic Ocean. The discovery made many things clearer, but also raised some new questions.

The *Titanic* had sunk in waters over 2.5 miles (4 km) deep. In 1912 there was no salvage equipment on earth able to go that deep, but by the 1980s there was. In 1981, a team financed by Texas oil magnate Jack Grimm searched the area around the location sent out by the *Titanic* as she sank with side-scan sonar and magnetometers, but with no success. Four years later a joint US–French team led by Jean Louis Michel and Dr Robert Ballard decided to try again with more sensitive sonar equipment. They mapped out an area of 400 square miles (1,036 km²) within which they thought the vagaries of ocean currents or mistaken positioning on the night might place the wreck. After three weeks the team had covered 80 per cent of the area without success.

On 1 September 1985 the team's underwater camera, mounted on a deep-sea probe named Argo, came across a huge metal object. It was a boiler from the *Titanic*. Next day another dive by the Argo found the bows of the *Titanic* sitting upright on the muddy seafloor in almost perfect condition. The team undertook a quick reconnaissance of the wreck before leaving for the winter. The following year the French pulled out for legal reasons and Ballard returned to undertake a more detailed study of the wreck.

The discovery of the *Titanic*'s wreck at once revealed two things. First, the ship had broken in two, and second, it was not where it was supposed to have been. The wreck lay at 41° 43' North 49° 56' West, about 13 miles (21 km) to the east and slightly south of where the distress signals had placed her sinking.

The front half of the ship was standing upright and more or less intact, though there was a gaping hole where the first funnel had fallen. The mast had fallen back on to the bridge. Just beside the position of the second funnel the ship ended. The stern was lying some 500 yd (457 m) away behind the front part, but turned around to face the other direction. If the front part was almost intact, the rear section was heavily damaged. Between the two sections and all around was a huge field of debris, such as plates, metal fittings and personal items.

It would seem that as the ship was sinking the front part of it filled up gradually with water. This dragged the front down, then tilted the stern up out of the water. As the stern reached an angle of about 20° or so, the weight of the metal suspended in the air became too great for the strength of the hull to bear. The hull seems to have cracked open beside the second funnel and behind the third funnel, allowing water to pour into the central section. The ship then lurched down and tipped even further towards the vertical. After a few minutes the bow section tore free and fell away. The stern, which was still filled with air, then bobbed back higher above the surface. The central section then tore free, probably causing the violent twisting motion to the stern remembered by survivors. After the central section fell away, the stern remained afloat and upright for as long as a minute. Then air began to escape and water to enter, allowing the stern to slide gently under the surface.

Once underwater the bow section levelled out as the water flowed over the pointed bows. It planed away with some forward motion, but falling downward only gently.

When it hit the muddy bottom, this section buried itself in the sediments but remained relatively intact. The central section, meanwhile, fell straight down and was smashed when it impacted on the bottom at speed. The rear section fared even worse. As it sank the increasing water pressure caused the pockets of air trapped within the hull to implode, causing massive damage. The section seems to have tumbled or spun as it fell and it hit the seafloor at speed, causing even more damage.

The newly discovered position of the wreck made sense of some of the puzzles left unresolved in 1912. Captain Moore of the *Mount Temple* took a clear dawn sight as he searched for survivors. He calculated that the position in the distress signal was just west of the ice field. Captain Lord of the *Californian* had made the same calculation, pushing through the ice field to reach where he estimated the lifeboats to be. But the *Carpathia* was picking up the survivors to the east of the ice field. As was now clear, the position sent out by the *Titanic* had been incorrect. Fourth Officer Boxhall had worked out the position as best he could, but he had made an error of some kind.

At last it was possible to put together a likely scenario for what had happened that night. *Titanic* was steaming at about 21.5 knots to the west. Ahead of her lay a large ice field made up principally of growlers and other low-lying field ice but interspersed with a few large icebergs, some of them very large. The *Californian* had met this ice field further north, but Lord had seen the starlight reflected off the expanse of field ice and come to a halt some 400 yd (365 m) from the ice.

The lookouts on the *Titanic* were keeping an eye open for exactly the same thing, but they never saw it. The iceberg they hit must have been drifting all alone some distance to the east of the main ice field, probably something approaching a mile and perhaps more. This ice was well beyond the usual southern extent of the drifting ice off Newfoundland and it was melting. The iceberg was seen by the lookout at a distance of less than 300 yd (275 m). It seems likely that an experienced and alert lookout would have sighted any iceberg at a much greater distance in normal conditions. The air was clear, so that had not caused the late sighting. More likely the iceberg was one that had recently calved or cracked in two in the melting waters. This meant it would have been one of the dreaded black icebergs that were so difficult to see at night.

Captain Lord is exonerated

What the revised position meant for the supporters of the deceased Captain Lord was that the *Titanic* had been even further away from Lord's estimated position than had been thought in 1912. Of course, that did not make Lord's estimate of his position any more accurate and he could still have been up to 5 miles (8 km) or more from where he thought he was. It did mean that it was almost impossible that the *Californian* had been within 3 miles (5 km) of the *Titanic* as Boxhall and others reported the mystery ship to have been. According to the traditional view of events, the mystery ship must have been the one also seen from the *Californian*. There had been a third ship right where Captain Lord had said it was.

It seems unlikely that either the schooner or the tramp steamer seen by Captain Moore of the *Mount Temple* could have been this ship. Both of them were on the western side of the ice field. The schooner can be ruled out as both those on the *Titanic* and those on the *Californian* reported seeing a steamer. The tramp steamer was, according to Moore, looking for a way through the ice and, having found one, steamed off to the east. The steamer seen by the *Californian* had come from the east and was heading west.

That leaves the steamer seen by Captain Rostron of the *Carpathia* as he was travelling north. If that steamer had been midway between the *Californian* and the *Titanic* and had left that position when Stone on the *Californian* reported that the ship he could see headed south, then it would have been passing the *Carpathia* some time between 3 am and 4 am. Rostron reported seeing it just after 3 am. Some witnesses on the *Titanic* said that the mystery ship left at soon after 2 am. On the other hand, Boxhall said it headed off to the west, and others thought it was still just north of the lifeboats as late as 3.30 am. If either are true it is unlikely to have been the ship that passed *Carpathia*.

There is, however, a third explanation for the various sightings that night. Several men mentioned this possibility but it does not seem to have been taken seriously at the time or since. Able Seaman John Poigndestre was definite that the light he saw was an optical illusion. He claimed it was a star seen though hazy air lying on the horizon, something he had seen before. Frenchman Omont also thought the lights were an optical illusion. His boat rowed hard to reach them, but they seemed to dance out of reach and then vanish. Finally Captain Lord said of visibility: 'It was a very deceiving night.'

The suggestion of an optical illusion seems to have been discounted because all witnesses said how perfectly clear and still the air was that night. But there is one type of rare optical illusion that does in fact take place in precisely such clear, still air conditions: the *fata morgana*. This occurs only when a band of warm air lies immediately on top of a band of much colder air and when the boundary between the two is calm and sharp. The air layers act like a lens, bending light rays down from the upper, warmer air into the colder, lower air. The upshot is that distant objects, sometimes those that are well over the horizon, appear to be larger and nearer than they really are. The shape of the object can also be distorted.

On the night the *Titanic* went down these conditions almost certainly existed. The field of ice chilled the sea around it and the air immediately above the sea – all witnesses who commented on the temperature reported this fact. But the general air in the North Atlantic was rather warmer and, moreover, was still. There were no winds to stir or mix the layers of air.

It should be remembered that nobody on the *Titanic* reported seeing the mystery ship itself, only its lights.

The reported activities of these lights – coming forward, fading away, getting brighter and then dimmer – are consistent with a *fata morgana*. It must be remembered that the men were desperate to see a rescue ship on the way. Perhaps they were seeing the lights of the *Californian* distorted through a *fata morgana*. There again, perhaps there really was a ship there.

The discovery of the wreck also raised the problem of who owned it. White Star had by then been merged with its old rival Cunard, but since the insurers had paid out for the loss they might be deemed to have ownership. Finally a US court awarded salvage rights to RMS *Titanic* Inc, an American company and subsidiary of the company that organized the expedition that found the remains. Since then the wreck has been exhaustively photographed inside and out, and some 6,000 objects recovered from the debris field. The legal debates over the *Titanic* have caused many countries to revise their laws on salvage and wreck recovery. The developing sophistication of undersea equipment means it is becoming increasingly feasible to raise objects from ancient or deep-sea wrecks.

The repeated visits to the liner by salvage vessels, remote controlled photographic probes and even tourist jaunts have been found to have an unexpected impact on the wreck. When it was first discovered in 1986 the ship had suffered only modest damage from rust, erosion and microbes. By 2010, however, the damage was found to be greatly advanced. As much damage had been done in the 24 years since discovery as in the 74 years the ship lay undisturbed. It is thought that the stirring of the waters and sediments by the probes and submersibles had caused the damage.

By 2100 the ship may have vanished entirely.

CHAPTER 12

THE SHIP
THAT KEEPS
ON SINKING

The dramatic events on board the *Titanic* have fascinated the world ever since they took place. The two inquiries between them established fairly clearly what had taken place and why it had happened. Numerous books and magazine articles written by survivors appeared, giving more of the human interest stories, anecdotes and incidents that fleshed out the bald facts and served to add emotional depth to the tragedy.

But the facts have not proved to be enough. The movie and entertainment business has produced a wealth of material about the *Titanic*. The ship sank only once in reality, but on the silver screen she has continued to sink again and again.

The first movie about the disaster was produced in Germany just six months after the ship foundered. Called *In Nacht und Eis* (In Night and Ice) the film was a form of docu-drama with actors playing the roles of the officers, crew and passengers. The script was based on the US inquiry, with Jack Phillips the radio operator being the main character. *In Nacht und Eis* ran for 35 minutes, about twice as long as an average movie of the time, and was considered an epic masterpiece in its day. Many scenes were shot on the 24,000-ton liner SS *Kaiserin Auguste Victoria* in Hamburg harbour with hoses being used to add water to simulate the incoming waters. The film also featured a 4-ft (1.2-m) model of the *Titanic* for the sinking scenes. By the 1920s all copies of the film appeared to have been lost and it was put down as a lost film – a movie of which no copy remains. However, a German collector found a copy in his collection in 1998, and a limited number of DVDs were released.

The Americans were next, releasing *Saved from the Titanic* a couple of weeks after *In Nacht und Eis*. The film starred Dorothy Gibson, the actress who had actually been on board the ship and escaped in Boat No. 7. In the movie Gibson wore the dress that she had been wearing on the night of the disaster. The movie was basically a reasonably faithful recreation of Miss Gibson's adventures as the ship went down, though a fictional love story was added and the movie ends with

Miss Gibson marrying her sweetheart in New York.

The following year the Danes produced another movie on the subject, *Atlantis*. The ship that sinks in the movie is a fictional liner called the *Roland*, but the parallels with events on the *Titanic* are so close that the two ships are clearly the same. The plot is actually a complex story about a Danish doctor emigrating to the New World and the sinking of the liner is only one incident in a long film of 113 minutes. The assistant director of the film, Michael Curtiz, later emigrated to the USA in reality where he became a successful director of movies such as *Casablanca* and *White Christmas*. The movie was considered to be a classic at its release and remains famous in Scandinavia. It is occasionally shown on television in Denmark and is screened at movie festivals from time to time.

In 1929 the British entered the field with a movie entitled *Atlantic*, the name of the fictional ship in the movie. The main focus of the film is the tangled emotional lives of a group of passengers. These include an elderly couple on a holiday to celebrate their long marriage, apparently based on the real life Straus couple on the *Titanic*. Their long-term relationship is contrasted with a younger couple who are recently married, but unhappy together. The younger husband embarks on a ship-board affair with a dissolute young woman. When the ship hits an iceberg the various characters are forced to confront their true feelings and resolve their assorted emotional dilemmas. The sinking of the ship is based very closely on the events on board the *Titanic* with the actions of Jack Phillips and Captain Smith in particular being shown reasonably accurately. The film runs to 87 minutes and was one of the first talkie movies made in Britain. The young female lead was taken by Madeleine Carroll, who went on to become one of the best-known British actresses of the 1930s and 1940s. She worked with Alfred Hitchcock in Britain, later moving to the US where she starred in the 1937 version of *The Prisoner of Zenda*.

The next appearance of the *Titanic* in a movie was in

a curious Nazi propaganda film released in Germany, occupied Europe and some neutral countries in 1943. The purpose of the movie, that was the first to be called simply *Titanic*, was to highlight the cruelty, corruption and arrogance of wealthy British and American businessmen. It also showed how their greed and lack of morals caused massive loss of life among working-class Britons. The intent was to demonstrate that the British and American involvement in the war was driven by similar motives. The British plutocrats – one of whom bears a physical resemblance to the then British Prime Minister Winston Churchill – are shown as a set of cruel and heartless oppressors. Down in the third-class cabins are a group of plucky German working-class folk who espouse Nazi doctrine. As the ship goes down, these Germans manage to outwit the cruel Brits and so save some deserving women and children. The audience is then treated to a view of the deaths of the oppressors in the icy waters. The movie also featured a fictional German officer who argues with Ismay over the ship's speed, warns about icebergs and is generally always right. Towards the end he persuades his girlfriend to get into a lifeboat by lying to her about his own place in a different boat.

The movie ends with the German officer appearing at the British inquest where he points the finger of blame at Ismay, making some propaganda points of German honesty contrasted with British greed and cruelty. The inquiry, however, clears Ismay and blames the dead Captain Smith, thus once again emphasizing British dishonesty.

The movie was the most expensive German film made up to that date. It was shot on board the 12,000-ton liner SS *Cap Arcona* in the Baltic Sea. No expense was spared on special effects or the cast, and it was a big hit both in Germany and in the other countries where it was shown. After the war it was recut to remove the more blatant pro-Nazi propaganda and re-released to critical acclaim and public enthusiasm. This version has been shown on German television several times over the years. In 2005 the original, uncut version was released on DVD.

Not to be outdone, Hollywood produced its own movie *Titanic* in 1953 starring Clifton Webb and Barbara Stanwyck. The storyline concerned a fictitious couple, an American woman and British man, who are currently going through a divorce. The woman is travelling back home to the USA with her two children, while the husband books himself on to the ship at the last moment in the hope of achieving a reconciliation. The tangled love story is finally resolved as the ship begins to sink when the couple pledge their renewed love, but are then divided again as the wife gets into a lifeboat and the husband remains on the ship to drown. Among the other passengers are many who really were on the ship, though some appear under fictitious names as they were still alive.

The movie proved to be popular, but soon became famous for its many inaccuracies. These ranged from the minor – having the wrong instruments in the ship's band and putting the crew in Royal Navy uniforms – to the more serious such as having the ship sink on the wrong date, giving the ship an evacuation alarm, having Captain Smith in command when the ship hit the iceberg and having an internal layout that was completely unlike that of the real ship. Despite these and dozens of other mistakes, the screenplay won an Oscar.

Five years later the British produced *A Night to Remember* starring Kenneth More. This film is essentially a docu-drama and is far and away the most accurate of all the film versions. More took the role of Second Officer Lightoller who is very much the hero of the film as he organizes the evacuation. The plot covers events from when the ship left Ireland to when the survivors are rescued by the *Carpathia*. The film was the first to show events on board the SS *Californian*, and depicts Captain Lord in a very poor light. This rekindled interest in Lord and his actions (he was still alive at the time) and led to research that convinced many that he had been badly treated both by the two inquiries and by this film.

The film shows numerous real life people, giving them the actions and words that they actually did say and do. It does a fair job of trying to condense the many and varied events of the night into a two-hour film and certainly gives a good impression of the changing atmosphere on board as the sinking progressed. There are, however, a few errors. The picture that the ship's designer Thomas Andrews stared at as the ship goes down is shown to be a view of New York, but in reality it was a view of Plymouth. The Socket distress signals are shown as conventional firework rockets. More seriously it shows the stewards keeping all third-class passengers down below when in fact great efforts were made to escort the women and children up to the lifeboats – not always successfully, it must be said. The film is out on DVD.

Blockbuster

The next movie to feature the ship was the 1997 blockbuster *Titanic* made by James Cameron and starring Leonardo DiCaprio and Kate Winslet. This film reverted to having a fictitious love story as the central plot. This time, Winslet plays a British woman from an impoverished aristocratic family on her way to America to be married to a rich, but snobbish and heartless American business magnate, played by Billy Zane. While on board she meets and falls in love with a penniless American artist, played by DiCaprio. The tortured love triangle unfolds as the ship steams across the Atlantic. Once the ship hits the iceberg, the two lovers realize the depth of their feelings and spend the rest of the film trying to escape both the sinking ship and the vengeful jilted fiancé.

Although the three fictional characters occupy most of the film, many real characters are also shown. All the officers are shown, along with the two radio men, Molly Brown, Colonel Gracie, Benjamin Guggenheim, the Straus couple, the Astors, Ismay, Andrews and many others. The film was the most expensive movie ever made, costing around $200 million all told. The producers built a full-sized replica of the ship for the outside shots, and made detailed reproductions of the interiors in the studio. The attention to detail is impressive; even the pattern of the plates that the actors eat off is correct. The special effects used in the sinking were complex and pushed what was possible to the limit. The film won 11 Oscars and was a huge box office success.

Despite the attention to detail, the film did make a few errors. Like *A Night to Remember* it shows the *Titanic* firing off distress rockets, not the specialized Socket signals. Again following *A Night to Remember*, it shows the third-class passengers being kept below to drown while the first-class passengers escape. This time the passengers are shown being kept behind locked gates, not just being advised to stay below by stewards. It adds drama to the film, but is entirely wrong.

The movie also missed out many of the events during the sinking. It made no mention of the mystery ship or of the *Californian*. No effort was made to explain the evacuation procedure with the process being depicted as something of a disorganized scrum. Although the officers are shown, most of them are hardly seen once the ship begins to sink apart from Captain Smith and First Officer Murdoch. When Lightoller is depicted he is shown getting close to panic, which is hardly fair on him.

The movie also shows First Officer Murdoch taking a bribe to allow a man into a boat – though he later throws the bribe back and refuses the man a place. Murdoch is also shown shooting himself, an incident for which there is no firm evidence.

Movies are, however, only the most high profile appearances of the *Titanic*. The ship has also featured in novels, magazines and on numerous television shows. It has appeared not only as itself, but also in different guises. In an episode of the BBC science fiction series Dr Who, there is an alien spaceship named the *Titanic* on which aliens enjoy a holiday as if they were on early 20th-century Earth. Inevitably, the spaceship *Titanic* is heading for disaster, though the Doctor manages to save it.

The *Titanic* has become so famous as a symbol of maritime disaster that it seems to be irresistible to movie makers and TV producers. But that should not detract from the very real tragedy of the real-life *Titanic*. Nowhere is that tragedy more tangible than in Halifax, Nova Scotia. It was here that the bodies of the dead recovered by the ship Mackay-Bennett sent out by the White Star Line were brought. Those families who wanted to or could afford the cost, were allowed to transport the bodies of their loved ones for burial in their home towns. The rest were buried at the Fairlawn Cemetery at Halifax. Most of the 333 bodies recovered were identified from belongings in their pockets – John Jacob Astor's body had thousands of dollars in his pockets and a fortune in gold and diamond jewellery. A few lacked any personal belongings and could not be identified.

Among the unidentified bodies was one that deeply upset the people of Halifax. It was the body of a cherubic boy with fair hair who was aged about 2 years old. The citizens collected enough money to give the boy a special tombstone on which was inscribed:

Erected to the memory of an unknown child whose

remains were recovered after the disaster to the Titanic April 15th 1912.

The tomb is still there. To this day it is regularly bedecked with fresh flowers and cuddly toys.

BIBLIOGRAPHY AND FURTHER READING

No discussion of the *Titanic* would be complete without referring to Walter Lord's classic book *A Night to Remember* (Bantam, 1955). This book drew on many eyewitnesses interviewed by Lord as well as on contemporary newspaper accounts and the two inquiries. It must be said, however, that the conclusions drawn by Lord are not beyond criticism, especially as regards his explanation of the mystery ship and the evacuation procedure on the *Titanic*. As the title suggests, the book concentrates exclusively on the events of the sinking and has little to say about the construction of the ship and other issues.

Charles Lightoller's book *Titanic and Other Ships* (Ivor Nicholson and Watson, 1936) makes enjoyable reading. It is largely an autobiographical work in which the sinking of the *Titanic* is but one among many events, but it gives a very good idea of what it was like to ply the oceans as an officer on passenger liners and provides a wealth of background information about shipboard routine and life that is essential reading to make sense of the events on the *Titanic*.

Archibald Gracie wrote **The Truth about the Titanic** (Mitchell Kennerley, 1913) just a few months after the tragedy – later republished as *The Titanic: A Survivor's Story* (N C Multimedia Corporation, 1986). It is a gripping read that gives a detailed account of Gracie's own experiences on the night and includes a wealth of anecdotes that Gracie collected from other survivors. However, Gracie's speculations and guesses about events that happened outside his own direct experience are often wide of the mark.

These three books are still in print and are easy to find. Other accounts by survivors are not in print and are more difficult to come by. The best of these are Harold Bride's

Titanic's Surviving Wireless Man and *The Loss of the SS Titanic* by first-class passenger Lawrence Beesley.

The British National Archives holds the complete transcripts of the British public inquiry, along with all the associated letters and other paperwork. This runs to several thousand pages of paper which is, fortunately, well archived and indexed as well as being available on microfilm. The documents can all be viewed by making an appointment via the contact details given on the website **www.nationalarchives.gov.uk**. The transcripts and documents of the US Senate Hearings are similarly available via the US National Archives, and details of how to view documents can be found on **www.archives.gov**.

All British, American, French and Scandinavian newspapers carried extensive reports on the sinking in the days that followed the tragedy. Most of these are available in national archives for viewing by appointment. Some newspapers maintain their own archives and may be willing to allow access under certain conditions.

Robert Ballard, who discovered the wreck, produced the lavishly illustrated **Return to Titanic** (National Geographic, 2004) which is an admirable visual and factual account of the wreck.

In addition to these source books there is a huge number of other books on the *Titanic*. These range from careful academic studies through to colouring books for children. Among the books written for the general reading public the following are worth reading:

Robin Gardiner and Dan Van Der Vat, *The Riddle Of The Titanic*, Weidenfeld & Nicolson, 1995.

Don Lynch, *Titanic: An Illustrated History*, Hyperion Books, 1998.

Rod Green, *Building the Titanic: An Epic Tale of Human Endeavour and Modern Engineering*, Carlton Books, 2005.

Jennifer McCarty, *What Really Sank The Titanic: New Forensic Discoveries*, Citadel Press, 2008.

Alan Scarth, *Titanic and Liverpool*, Liverpool University Press, 2009.